SpringerBriefs in Mathematics

Series Editors

Nicola Bellomo, Torino, Italy

Michele Benzi, Pisa, Italy

Palle Jorgensen, Iowa City, USA

Tatsien Li, Shanghai, China

Roderick Melnik, Waterloo, Canada

Otmar Scherzer, Linz, Austria

Benjamin Steinberg, New York, NY, USA

Lothar Reichel, Kent, USA

Yuri Tschinkel, New York, NY, USA

George Yin, Detroit, USA

Ping Zhang, Kalamazoo, MI, USA

SpringerBriefs present concise summaries of cutting-edge research and practical applications across a wide spectrum of fields. Featuring compact volumes of 50 to 125 pages, the series covers a range of content from professional to academic. Briefs are characterized by fast, global electronic dissemination, standard publishing contracts, standardized manuscript preparation and formatting guidelines, and expedited production schedules.

Typical topics might include:

- A timely report of state-of-the art techniques
- A bridge between new research results, as published in journal articles, and a contextual literature review
- A snapshot of a hot or emerging topic
- An in-depth case study
- A presentation of core concepts that students must understand in order to make independent contributions

SpringerBriefs in Mathematics showcases expositions in all areas of mathematics and applied mathematics. Manuscripts presenting new results or a single new result in a classical field, new field, or an emerging topic, applications, or bridges between new results and already published works, are encouraged. The series is intended for mathematicians and applied mathematicians. All works are peer-reviewed to meet the highest standards of scientific literature.

Titles from this series are indexed by Scopus, Web of Science, Mathematical Reviews, and zbMATH.

Alexander J. Zaslavski

Turnpike Phenomenon for Markov Decision Processes

 Springer

Alexander J. Zaslavski
Department of Mathematics
Technion, Israel Institute of Technology
Haifa, Israel

ISSN 2191-8198 ISSN 2191-8201 (electronic)
SpringerBriefs in Mathematics
ISBN 978-3-032-00853-4 ISBN 978-3-032-00854-1 (eBook)
https://doi.org/10.1007/978-3-032-00854-1

Mathematics Subject Classification: 90C40

This Springer imprint is published by the registered company Springer Nature Switzerland AG
The registered company address is: Gewerbestrasse 11, 6330 Cham, Switzerland

If disposing of this product, please recycle the paper.

Preface

The monograph is devoted to the study of the structure of approximate optimal policies of a Markov decision process with a finite set of states and approximate optimal solutions of corresponding general deterministic discrete-time optimal control problems. It contains new results on properties of approximate solutions (policies) which are independent of the length of the interval, for all sufficiently large intervals. These results deal with the so-called turnpike property of optimal control problems. The term was first coined by P. Samuelson in 1948 when he showed that an efficient expanding economy would spend most of the time in the vicinity of a balanced equilibrium path (also called a von Neumann path). To have the turnpike property means, roughly speaking, that the approximate solutions of the problems are determined mainly by the objective function and are essentially independent of the choice of interval and endpoint conditions, except in regions close to the endpoints. Now it is well-known that the turnpike property is a general phenomenon which holds for large classes of variational and optimal control problems. Using the Baire category (generic) approach, it was shown that the turnpike property holds for a generic (typical) variational problem [93] and for a generic optimal control problem [99].

In Chap. 1 the turnpike phenomenon is explained and the structure of the book is described. In Chap. 2 we study infinite horizon discrete-time optimal control of Markov Decision Processes (MDPs) with finite state spaces and compact action sets and employ the long-run expected average cost criterion. We study the uniqueness and stability of minimizing Markov actions and show that in a set of MDPs equipped with a natural complete metric subset of MDPs with essentially unique and stable minimizing Markov actions contains a countable intersection of open everywhere dense sets. Thus, the property of having essentially unique and stable minimizing Markov actions is generic for this class of MDPs. In Chap. 3 we consider a subclass of MDPs studied in Chap. 2 and assuming the uniqueness of minimizing Markov actions show the existence of an overtaking optimal policy. We also establish the asymptotic turnpike property for good programs on infinite horizon. Chapter 4 contains the study of the class of MDPs considered in Chap. 3. Assuming the uniqueness of minimizing Markov actions we establish the weak turnpike property for approximate optimal programs on finite horizon. In Chap. 5 we consider the subclass of MDPs studied in Chap. 2 with convex action sets and convex cost functions. We show that if cost functions are strictly convex, then minimizing Markov actions are unique and that most in the sense of Baire category convex cost functions are in fact strictly convex. Chapter 6 is devoted to the study of turnpike properties for MDPs with perturbations. We assume the uniqueness of minimizing Markov actions and show that the asymptotic turnpike property and the weak turnpike property are stable under the perturbations of cost functions. In Chaps. 2–6 the study of our stochastic optimal control problem is reduced to the study of the corresponding deterministic optimal control problem which is a particular case of a general optimal control system which will be studied in Chap. 8. Turnpike properties for deterministic optimal control systems are well-known in the literature but they are related to states of the system. But for our MDPs and the related deterministic optimal control system the turnpike phenomenon holds both for its states and its controls. Therefore the existing turnpike results cannot be applied to our situation and new turnpike results are needed. They are presented in Chap. 8. As usual our strong turnpike property will be deduced from the asymptotic turnpike property and some controllability conditions which are discussed in Chap. 7.

Rishon LeZion, Israel   Alexander J. Zaslavski
January 15, 2025

Declarations

Competing Interests The author has no competing interests to declare that are relevant to the content of this manuscript.

Contents

Chapter 1
Introduction

The study of the existence and the structure of solutions of optimal control problems defined on infinite intervals and on sufficiently large intervals has recently been a rapidly growing area of research. See, for example, [7, 8, 11, 12, 16, 18–20, 26, 27, 35, 37, 38, 46, 51, 60, 64, 74, 93, 98, 99, 101], and the references mentioned therein. These problems arise in engineering [26, 93], in models of economic growth [6, 24, 25, 36, 59, 69, 82, 86, 93, 100], in infinite discrete models of solid-state physics related to dislocations in one-dimensional crystals [10, 83, 92], in the theory of thermodynamical equilibrium for materials [67, 70–72] and in optimal control theory with PDEs [39–41, 79, 90, 102]. In this chapter we discuss the structure of solutions of a discrete-time optimal control system describing a general model of economic dynamics studied in [94, 96, 97]. This discussion illustrates the meaning of the turnpike phenomenon. We also describe the structure of the book.

1.1 The Turnpike Phenomenon

We study the structure of approximate solutions of an autonomous discrete-time control system with a compact metric space of states X equipped with a metric ρ. This control system is described by a bounded upper semicontinuous function $v : X \times X \to R^1$ which determines an optimality criterion and by a nonempty closed set $\Omega \subset X \times X$ which determines a class of admissible trajectories (programs). In models of economic growth the set X is the space of states, v is a utility function and $v(x_t, x_{t+1})$ evaluates consumption at moment t.

Note that for this control system for every instant of time t the control variable satisfies $u_t = x_{t+1}$.

A. J. Zaslavski, *Turnpike Phenomenon for Markov Decision Processes*, SpringerBriefs in Mathematics, https://doi.org/10.1007/978-3-032-00854-1_1

Consider the problem

$$\sum_{i=0}^{T-1} v(x_i, x_{i+1}) \to \max, \quad \{(x_i, x_{i+1})\}_{i=0}^{T-1} \subset \Omega, \ x_0 = z, \ x_T = y, \qquad (P)$$

where $T \geq 1$ is an integer and the points $y, z \in X$. Note that for this control system for every instant of time t the control variable satisfies $u_t = x_{t+1}$.

We are interested in the turnpike property of approximate solutions which is independent of the length of the interval, for all sufficiently large intervals. To have this property means, roughly speaking, that approximate solutions of optimal control problems on an interval $[0, T]$ with given values y, z at the endpoints 0 and T, corresponding to the pair (v, Ω), are determined mainly by the objective function v, and are essentially independent of T, y and z.

In the classical turnpike theory the objective function v possesses the turnpike property (TP) if there exists a point $\bar{x} \in X$ (a turnpike) such that the following condition holds:

For each positive number ϵ there exists an integer $L \geq 1$ such that for each integer $T \geq 2L$ and each solution $\{x_i\}_{i=0}^{T} \subset X$ of the problem (P) the inequality $\rho(x_i, \bar{x}) \leq \epsilon$ is true for all $i = L, \ldots, T - L$.

It should be mentioned that the constant L depends neither on T nor on y, z.

The turnpike phenomenon has the following interpretation. If one wishes to reach a point A from a point B by a car in an optimal way, then one should turn to a turnpike, spend most of time on it and then leave the turnpike to reach the required point.

In the classical turnpike theory [36, 69, 73] the space X is a compact convex subset of a finite-dimensional Euclidean space, the set Ω is convex and the function v is strictly concave. Under these assumptions the turnpike property can be established and the turnpike $\bar{x}$ is a unique solution of the maximization problem $v(x, x) \to \max, (x, x) \in \Omega$. In this situation it is shown that for each admissible sequence $\{x_t\}_{t=0}^{\infty}$ either the sequence $\{\sum_{t=0}^{T-1} v(x_t, x_{t+1}) - T v(\bar{x}, \bar{x})\}_{T=1}^{\infty}$ is bounded (in this case the sequence $\{x_t\}_{t=0}^{\infty}$ is called (v)-good) or it diverges to $-\infty$. Moreover, it is also established that any (v)-good admissible sequence converges to the turnpike $\bar{x}$. In the sequel this property is called as the asymptotic turnpike property.

Now it is well-known that the turnpike property is a general phenomenon which holds for large classes of variational and optimal control problems without convexity assumptions. (See, for example, [93, 98, 99] and the references mentioned therein.) For these classes of problems a turnpike is not necessarily a singleton but may instead be an nonstationary trajectory (in the discrete time nonautonomous case) or an absolutely continuous function on the interval $[0, \infty)$ (in the continuous time nonautonomous case) or a compact subset of the space X (in the autonomous case).

For classes of problems considered in [93, 99], using the Baire category approach, it was shown that the turnpike property holds for a generic (typical) problem. In this book we are interested in individual (non-generic) turnpike results and in stability of the turnpike phenomenon under small perturbations of an objective function and a control set.

As we have mentioned before in general a turnpike is not necessarily a singleton. Nevertheless problems of the type (P) for which the turnpike is a singleton are of great importance because of the following reasons: there are many models in applications for which a turnpike is a singleton; if a turnpike is a singleton, then approximate solutions of (P) have very simple structure and this is very important for applications; if a turnpike is a singleton, then it can be easily calculated.

The turnpike property is very important for applications. Suppose that our objective function v has the turnpike property and we know a finite number of "approximate" solutions of the problem (P). Then we know the turnpike $\bar{x}$, or at least its approximation, and the constant L (see the definition of (TP)) which is an estimate for the time period required to reach the turnpike. This information can be useful if we need to find an "approximate" solution of the problem (P) with a new time interval $[m_1, m_2]$ and the new values $z, y \in X$ at the end points m_1 and m_2. Namely instead of solving this new problem on the "large" interval $[m_1, m_2]$ we can find an "approximate" solution of the problem (P) on the "small" interval $[m_1, m_1 + L]$ with the values $z, \bar{x}$ at the end points and an "approximate" solution of the problem (P) on the "small" interval $[m_2 - L, m_2]$ with the values $\bar{x}, y$ at the end points. Then the concatenation of the first solution, the constant sequence $x_i = \bar{x},\ i = m_1 + L, \ldots, m_2 - L$ and the second solution is an "approximate" solution of the problem (P) on the interval $[m_1, m_2]$ with the values z, y at the end points. Sometimes as an "approximate" solution of the problem (P) we can choose any admissible sequence $\{x_i\}_{i=m_1}^{m_2}$ satisfying

$$x_{m_1} = z,\ x_{m_2} = y \text{ and } x_i = \bar{x} \text{ for all } i = m_1 + L, \ldots, m_2 - L.$$

In the sequel we denote by $\mathrm{Card}(A)$ the cardinality of a set A. We suppose that the sum over empty set is zero.

1.2 Discrete-Time Problems

Let (X, ρ) be a compact metric space, Ω be a nonempty closed subset of $X \times X$ and let $v : X \times X \to R^1$ be a bounded upper semicontinuous function.

A sequence $\{x_t\}_{t=0}^{\infty} \subset X$ is called program if $(x_t, x_{t+1}) \in \Omega$ for all nonnegative integers t. A sequence $\{x_t\}_{t=0}^{T}$ where $T \geq 1$ is an integer is called a program if $(x_t, x_{t+1}) \in \Omega$ for all integers $t \in [0, T - 1]$.

We consider the problems

$$\sum_{i=0}^{T-1} v(x_i, x_{i+1}) \to \max,\ \{(x_i, x_{i+1})\}_{i=0}^{T-1} \subset \Omega,\ x_0 = y,$$

and

$$\sum_{i=0}^{T-1} v(x_i, x_{i+1}) \to \max, \ \{(x_i, x_{i+1})\}_{i=0}^{T-1} \subset \Omega, \ x_0 = y, \ x_T = z,$$

where $T \geq 1$ is an integer and the points $y, z \in X$.

We suppose that there exist a point $\bar{x} \in X$ and a positive number $\bar{c}$ such that the following assumptions hold:

(i) $(\bar{x}, \bar{x})$ is an interior point of Ω and the function v is continuous at the point $(\bar{x}, \bar{x})$;

(ii) $\sum_{t=0}^{T-1} v(x_t, x_{t+1}) \leq T v(\bar{x}, \bar{x}) + \bar{c}$ for any natural number T and any program $\{x_t\}_{t=0}^{T}$.

The property (ii) implies that for each program $\{x_t\}_{t=0}^{\infty}$ either the sequence

$$\{\sum_{t=0}^{T-1} v(x_t, x_{t+1}) - T v(\bar{x}, \bar{x})\}_{T=1}^{\infty}$$

is bounded or $\lim_{T \to \infty} [\sum_{t=0}^{T-1} v(x_t, x_{t+1}) - T v(\bar{x}, \bar{x})] = -\infty$.

A program $\{x_t\}_{t=0}^{\infty}$ is called (v)-good if the sequence

$$\{\sum_{t=0}^{T-1} v(x_t, x_{t+1}) - T v(\bar{x}, \bar{x})\}_{T=1}^{\infty}$$

is bounded. This notion was introduced in mathematical economics in [36].

Suppose that the following assumption holds.

(iii) (the asymptotic turnpike property) For any (v)-good program $\{x_t\}_{t=0}^{\infty}$, $\lim_{t \to \infty} \rho(x_t, \bar{x}) = 0$.

Note that the properties (i)–(iii) hold for models of economic dynamics considered in the classical turnpike theory.

For each positive number M denote by X_M the set of all points $x \in X$ for which there exists a program $\{x_t\}_{t=0}^{\infty}$ such that $x_0 = x$ and that for all natural numbers T the following inequality holds:

$$\sum_{t=0}^{T-1} v(x_t, x_{t+1}) - T v(\bar{x}, \bar{x}) \geq -M.$$

It is not difficult to see that $\cup \{X_M : M \in (0, \infty)\}$ is the set of all points $x \in X$ for which there exists a (v)-good program $\{x_t\}_{t=0}^{\infty}$ satisfying $x_0 = x$.

Let $u : X \times X \to R^1$, be a bounded function, an integer $T \geq 1$ and $\Delta \geq 0$. A program $\{x_i\}_{i=0}^{T} \subset X$ is called (u, Δ)-optimal if for any program $\{x_i'\}_{i=0}^{T}$ satisfying

$x_0 = x'_0$, the inequality

$$\sum_{i=0}^{T-1} u(x_i, x_{i+1}) \geq \sum_{i=0}^{T-1} u(x'_i, x'_{i+1}) - \Delta$$

holds.

The following turnpike result describes the structure of approximate solutions of our first optimization problem stated above.

Theorem 1.1 *Let ϵ, M be positive numbers. Then there exist a natural number L and a positive number δ such that for each integer $T > 2L$ and each (v, δ)-optimal program $\{x_t\}_{t=0}^{T}$ which satisfies $x_0 \in X_M$ there exist nonnegative integers $\tau_1, \tau_2 \leq L$ such that $\rho(x_t, \bar{x}) \leq \epsilon$ for all $t = \tau_1, \ldots, T - \tau_2$ and if $\rho(x_0, \bar{x}) \leq \delta$, then $\tau_1 = 0$.*

An analogous turnpike result also holds for approximate solutions of our second optimization problem.

A program $\{x_t\}_{t=0}^{\infty}$ is called (v)-overtaking optimal if for each program $\{y_t\}_{t=0}^{\infty}$ satisfying $y_0 = x_0$ the inequality

$$\limsup_{T \to \infty} \sum_{t=0}^{T-1} [v(y_t, y_{t+1}) - v(x_t, x_{t+1})] \leq 0$$

holds. This notion was introduced in [91]. See also [85].

The following result establishes the existence of an overtaking optimal program.

Theorem 1.2 *Assume that $x \in X$ and that there exists a (v)-good program $\{x_t\}_{t=0}^{\infty}$ such that $x_0 = x$. Then there exists a (v)-overtaking optimal program $\{x_t^*\}_{t=0}^{\infty}$ such that $x_0^* = x$.*

The following result provides necessary and sufficient conditions for overtaking optimality.

Theorem 1.3 *Let $\{x_t\}_{t=0}^{\infty}$ be a program such that*

$$x_0 \in \cup\{X_M : M \in (0, \infty)\}.$$

Then the program $\{x_t\}_{t=0}^{\infty}$ is (v)-overtaking optimal if and only if the following conditions hold:

(1) $\lim_{t \to \infty} \rho(x_t, \bar{x}) = 0$;
(2) for each natural number T and each program $\{y_t\}_{t=0}^{T}$ satisfying $y_0 = x_0$, $y_T = x_T$ the inequality $\sum_{t=0}^{T-1} v(y_t, y_{t+1}) \leq \sum_{t=0}^{T-1} v(x_t, x_{t+1})$ holds.

The next two theorems establish uniform convergence of overtaking optimal programs to $\bar{x}$.

Theorem 1.4 *Assume that the function v is continuous and let ϵ be a positive number. Then there exists a positive number δ such that for each (v)-overtaking optimal program $\{x_t\}_{t=0}^{\infty}$ satisfying $\rho(x_0, \bar{x}) \leq \delta$ the inequality $\rho(x_t, \bar{x}) \leq \epsilon$ holds for all nonnegative integers t.*

Theorem 1.5 *Assume that the function v is continuous and let M, ϵ be positive numbers. Then there exists an integer $L \geq 1$ such that for each (v)-overtaking optimal program $\{x_t\}_{t=0}^{\infty}$ satisfying $x_0 \in X_M$ the inequality $\rho(x_t, \bar{x}) \leq \epsilon$ holds for all integers $t \geq L$.*

Theorems 1.1–1.5 were obtained in [94].

Example 1.6 Let (X, ρ) be a compact metric space, Ω be a nonempty closed subset of $X \times X$, $\bar{x} \in X$, $(\bar{x}, \bar{x})$ be an interior point of Ω, $\pi : X \to R^1$ be a continuous function, α be a real number and $L : X \times X \to [0, \infty)$ be a continuous function such that for each $(x, y) \in X \times X$ the equality $L(x, y) = 0$ holds if and only if $(x, y) = (\bar{x}, \bar{x})$. Set

$$v(x, y) = \alpha - L(x, y) + \pi(x) - \pi(y)$$

for all $x, y \in X$. It is not difficult to see that assumptions (i), (ii) and (iii) hold.

Example 1.7 Let X be a compact convex subset of the Euclidean space R^n with the norm $|\cdot|$ induced by the scalar product $\langle \cdot, \cdot \rangle$, let $\rho(x, y) = |x - y|$, $x, y \in R^n$, Ω be a nonempty closed subset of $X \times X$, a point $\bar{x} \in X$, $(\bar{x}, \bar{x})$ be an interior point of Ω and let $v : X \times X \to R^1$ be a strictly concave continuous function such that

$$v(\bar{x}, \bar{x}) = \sup\{v(z, z) : \ z \in X \text{ and } (z, z) \in \Omega\}.$$

We assume that there exists a positive constant $\bar{r}$ such that

$$\{(x, y) \in R^n \times R^n : \ |x - \bar{x}|, \ |y - \bar{x}| \leq \bar{r}\} \subset \Omega.$$

It is a well-known fact of convex analysis [84] that there exists a point $l \in R^n$ such that

$$v(x, y) \leq v(\bar{x}, \bar{x}) + \langle l, x - y \rangle$$

for any point $(x, y) \in X \times X$. Set

$$L(x, y) = v(\bar{x}, \bar{x}) + \langle l, x - y \rangle - v(x, y)$$

for all $(x, y) \in X \times X$. It is not difficult to see that this example is a particular case of Example 1.6. Therefore assumptions (i), (ii) and (iii) hold.

Denote by $\mathcal{M}$ the set of all bounded functions $u : X \times X \to R^1$. For each function $w \in \mathcal{M}$ we set

$$\|w\| = \sup\{|w(x, y)| : x, y \in X\}.$$

The following two theorems obtained in [96, 97] respectively show that the turnpike property is stable under perturbations of the objective function. The first result establishes the strong turnpike property while the second one demonstrates its weak version.

Theorem 1.8 *Let M_0, $\epsilon > 0$. Then there exist a positive number δ and an integer $L_* \geq 1$ such that for each function $u \in \mathcal{M}$ satisfying $\|u - v\| \leq \delta$, each integer $T > 2L_*$ and each (u, δ)-optimal program $\{x_t\}_{t=0}^{T}$ which satisfies $x_0 \in X_{M_0}$ there exist integers $\tau_1 \in [0, L_*]$, $\tau_2 \in [T - L_*, T]$ such that*

$$\rho(x_t, \bar{x}) \leq \epsilon, \ t = \tau_1, \ldots, \tau_2.$$

Moreover if $\rho(x_0, \bar{x}) \leq \delta$, then $\tau_1 = 0$.

Theorem 1.9 *Let M_0, M_1, ϵ be positive numbers. Then there exist a positive number δ and an integer $L_* \geq 1$ such that for each function $u \in \mathcal{M}$ satisfying $\|u - v\| \leq \delta$, each integer $T > L_*$ and each (u, M_1)-optimal program $\{x_t\}_{t=0}^{T}$ which satisfies $x_0 \in X_{M_0}$ the following inequality holds:*

$$Card(\{t \in \{0, \ldots, T\} : \ \rho(x_t, \bar{x}) > \epsilon\}) \leq L_*.$$

Note the for models of economic dynamics the first weak turnpike result was obtained by R. Radner [81] while the strong turnpike property was firstly studied by H. Nikaido [75].

1.3 Examples

Example 1.10 Let $X = [0, 1]$, $\Omega = \{(x, y) \in [0, 1] \times [0, 1] : \ y \leq x^{1/2}\}$, $v(x, y) = x^{1/2} - y^2, x, y \in X$. It is not difficult to see that the set Ω is convex, the function v is strictly concave, the optimization problem $v(z, z) \to \max, z \in X$ and $(z, z) \in \Omega$ has a unique solution $16^{-1/3}$ and $(16^{-1/3}, 16^{-1/3})$ is an interior point of Ω. Therefore this example is a particular case of Example 1.7 and assumptions (i), (ii) and (iii) hold.

Example 1.11 Let $X = [0, 1]$, $\Omega = \{(x, y) \in [0, 1] \times [0, 1] : \ y \leq x^{1/2}\}$, $v(x, y) = x^{1/2} - y, x, y \in X$. It is not difficult to see that the set Ω is convex, the function v is concave but not strictly concave, the optimization problem $v(z, z) \to \max, z \in X$ and $(z, z) \in \Omega$ has a unique solution 4^{-1} and $(4^{-1}, 4^{-1})$ is an interior point of Ω. Since the function v is concave for all $x, y \in X$,

$$v(x, y) \le v(4^{-1}, 4^{-1}) + x - y = 4^{-1} + x - y$$

and

$$4^{-1} + x - y - v(x, y) = (x^{1/2} - 2^{-1})^2$$

is equal zero if and only if $x = 4^{-1}$. Now it is not difficult to see that assumptions (i), (ii) and (iii) hold.

Example 1.12 Consider the sets X, Ω and the function v defined in Example 1.11 and set $u(x, y) = x^{1/2} - x^2 - y + y^2$, $x, y \in X$. The function u is strictly convex with respect to the variable y. Nevertheless assumptions (i), (ii) and (iii) hold for the function u because for any integer T and any program $\{x_t\}_{t=0}^{T}$,

$$\sum_{t=0}^{T-1} u(x_t, x_{t+1}) = \sum_{t=0}^{T-1} v(x_t, x_{t+1}) + x_T^2 - x_0^2.$$

1.4   Markov Decision Processes

Infinite horizon stochastic optimal control has being a rapidly growing area of research [2, 3, 5, 30, 31, 42, 45, 49, 50, 52, 53, 55, 57, 61–63, 80, 87] and the study of Markov Decision Processes (MDPs) is one of its central topics [4, 13, 15, 17, 21–23, 28, 29, 32–34, 43, 44, 47, 54, 56, 65, 66, 68, 76–78, 88]. The first research on the turnpike theory for finite-state MDPs was done in [89].

In this book we study infinite horizon discrete-time optimal control of Markov Decision Processes with finite state spaces and compact action sets, and employ the long-run expected average cost criterion. MDPs provide a model for controlled dynamic processes, where the controller's goal is to minimize a specified cost function. The model is defined in terms of certain parameters, in general these cannot be determined exactly, and usually there are some uncertainties concerning their values.

In this book we use the following notions. We denote by $X = \{1, \ldots, m\}$ the state space, thus at each instant of time $k \ge 0$ the controlled system occupies a state $\xi_k = i \in X$. Choosing an action a at this time has the effect of shifting the state to $\xi_{k+1} = j \in X$ at time $k + 1$, with various transition probabilities $Q(j|i, a)$ which depend on the chosen action a. We suppose that the set of actions is a subset of

$$A = \{(a_1, \ldots, a_m) \in R^m : a_i \ge 0, \ i = 1, \ldots m, \text{ and } \sum_{i=1}^{m} a_i = 1\}$$

(see [13–15, 32]). Thus in this approach one indicates directly the consequence of a choice of an action $a \in A$, and this by stating what are the probabilities a_i of transitions to the various states $i = 1, \ldots, m$.

Let $A(i)$ be the set of actions available when the process occupies the state i, and suppose that $A(i) \subset A$ is closed for every $i = 1, \ldots, m$.

Given two complete separable metric spaces X and Y we denote by $P(Y|X)$ the set of all stochastic kernels $\psi(dy|x)$ on Y given X, namely $\psi(dy|x)$ is a probability measure on Y for every fixed $x \in X$, and $\psi(B|\cdot)$ is a Borel measurable function on X for every Borel measurable set $B \subset Y$.

For vectors in R^m subscripts denote components and superscripts denote enumeration. Thus for a sequence $\{x^{(k)}\}_{k=1}^{\infty}$, $x_i^{(k)}$ is the i component of the k member of the sequence.

We next define the admissible control policies. For every $k \geq 0$ let H_k denote the set of histories h_k of the process up to epoch k:

$$h_k = (x^{(0)}) \text{ if } k = 0 \text{ and } h_k = (x^{(0)}, a^{(0)}, \ldots, x^{(k-1)}, a^{(k-1)}, x^{(k)}) \text{ if } k \geq 1,$$

with $a^{(i)} \in A(x^{(i)})$ for every integer i satisfying $0 \leq i \leq k - 1$. An admissible policy is a sequence $\{\sigma_0, \sigma_1, \sigma_2, \ldots\}$, where σ_k is a stochastic kernel on A given H_k (namely $\sigma_k \in P(A|H_k)$) with the property $\sigma_k(A(x^{(k)})|h_k) = 1$ for every integer $k \geq 0$ and every $h_k \in H_k$. Thus $\sigma_k(da|x^{(0)}, a^{(0)}, \ldots, x^{(k)})$ consists of a randomized rule for choosing an action $a^{(k)}$ on the basis of the information h_k. An admissible policy σ and an initial distribution $x^{(0)}$ determine a measure $P_{x^{(0)}}^{(\sigma)}$ on the set of sequences of elements in X. (See [52] for the construction of this measure.)

A policy is called Markovian if $\sigma(da|x^{(0)}, a^{(0)}, \ldots, x^{(k)}) = \tilde{\sigma}_k(da|x^{(k)})$ for some stochastic kernel $\tilde{\sigma}_k \in P(A(x^{(k)})|x^{(k)})$, $k \geq 0$ is an integer. A Markovian policy σ is called stationary Markovian if $\sigma_k(da|x^{(k)}) = \sigma_0(da|x^{(k)})$ for every integer $k \geq 0$, and it is called stationary and nonrandomized if the measure $\sigma_0(da|x)$ is concentrated in one point in $A(x)$ for every $x \in X$. For a Markovian policy σ the corresponding measure $P_{x^{(0)}}^{(\sigma)}$ is a measure of a Markov process on X.

For an $m \times m$ matrix M we denote by M_i the ith row of M. We denote by $\mathcal{M}$ the set of all matrices $M = (M_{ij})$, $i, j \in \{1, \ldots, m\}$, such that $M_i \in A(i)$ for every $i \in \{1, \ldots, m\}$. It follows that $M \in \mathcal{M}$ is a row stochastic matrix, which is the transition matrix of a stationary Markov chain on X, corresponding to the choice of the actions $M_i \in A(i)$. Thus there is 1-1 correspondence between stationary Markov policies and matrices M in $\mathcal{M}$, and we will henceforth represent a Markovian stationary policy σ by its corresponding matrix M as $\sigma = \{M^{\infty}\}$. More generally, in our framework a Markov policy is a sequence $\{M^{(k)}\}_{k=0}^{\infty}$, where $M^{(k)} \in \mathcal{M}$ for every $k \geq 0$.

We denote by $\| \cdot \|$ the Euclidean norm in R^m, and by $u \cdot v$ the scalar product of two vectors $u, v \in R^m$. We also define

$$|x|_{\infty} = \max\{|x_i| : i \in 1, \ldots, m\}, \quad x = (x_1, \ldots, x_m) \in R^n.$$

Put

$$G = \{(i, a) : i \in \{1, \ldots, m\}, \ a \in A(i)\}.$$

We consider the topological subspace $G \subset \{1, \ldots m\} \times A$ with the relative topology. Denote by $C(G)$ the space of all continuous functions $r : G \to R^1$ endowed with the norm

$$\|r\| = \max\{|r(i, a)| : \ (i, a) \in G\}, \ r \in C(G).$$

Clearly, $(C(G), \| \cdot \|)$ is a Banach space. Let $x^{(0)} \in A$ and let $\sigma = \{M^{(k)}\}_{k=0}^{\infty} \subset \mathcal{M}$ be a Markov policy. For the prescribed initial distribution $x^{(0)}$ we denote by $\{\xi^{(k)}\}_{k=0}^{\infty}$ and $\{a^{(k)}\}_{k=0}^{\infty}$ respectively the state and action processes generated by the policy σ. Thus $\{\xi^{(k)}\}_{k=0}^{\infty}$ is a Markov chain associated with $(x^{(0)}, \sigma)$:

$$P(\xi^{(0)} = i) = x_i^{(0)}, \ i = 1, \ldots m.$$

The conditional distribution of $\xi^{(k+1)}$ given that $\xi^{(k)} = j$ is given by the row vector $M_j^{(k)}$, namely the jth row of $M^{(k)}$:

$$P(\xi^{(k+1)} = l | \xi^{(k)} = j) = M_{jl}^{(k)}.$$

We denote by $x^{(k)}$ the distribution of $\xi^{(k)}$ and thus for each $l \in \{1, \ldots, m\}$,

$$P(\xi^{(k+1)} = l) = \sum_{j=1}^{m} x_j^{(k)} M_{j,l}^{(k)} = ((M^{(k)})^T x^{(k)})_l$$

and we have the relation: $x^{(k+1)} = (M^{(k)})^T x^{(k)}$ (where M^T is the transpose of M).

Let $r \in C(G)$ and $N \geq 1$ be an integer, and we consider a Markov policy $\sigma = \{M^{(k)}\}_{k=0}^{\infty}$. The state and action at time k are $\xi^{(k)}$ and $M_{\xi^{(k)}}^{(k)}$ respectively, so that the expected cost over the $[0, N]$ time interval is

$$C_N^r(x^{(0)}, \sigma) = E_{x^{(0)}}^{\sigma} \sum_{k=0}^{N-1} r(\xi^{(k)}, M_{\xi^{(k)}}^{(k)}), \tag{1.1}$$

where $E_{x^{(0)}}^{\sigma}$ is the expectation operator corresponding to the measure $P_{x^{(0)}}^{(\sigma)}$, which is associated with $(\sigma, x^{(0)})$. By induction we show that

$$C_N^r(x^{(0)}, \sigma) = \sum_{k=0}^{N-1} \sum_{j=1}^{m} x_j^{(k)} r(j, M_j^{(k)}). \tag{1.2}$$

Let $N = 1$. Then

$$C_1^r(x^{(0)}, \sigma) = E_{x^{(0)}}^\sigma r(\xi^{(0)}, M_{\xi^{(0)}}^{(0)}) = \sum_{j=1}^m x_j^{(0)} r(j, M_j^{(0)}).$$

Assume that $N \geq 1$ is an integer and (1.2) holds. Then we have

$$C_{N+1}^r(x^{(0)}, \sigma) = C_N^r(x^{(0)}, \sigma) + E_{x^{(0)}}^\sigma r(\xi^{(k+1)}, M_{\xi^{(k+1)}}^{(k+1)})$$

$$= C_N^r(x^{(0)}, \sigma) + \sum_{j=1}^m x_j^{(N)} r(j, M_j^{(N+1)})$$

and (1.2) holds for $N + 1$ too. Thus we showed by induction that (1.2) holds for every natural number N.

Chapter 2 of our book is devoted to the analysis of infinite horizon discrete-time optimal control of Markov Decision Processes (MDPs) with finite state spaces and compact action sets. For these MDPs we employ the long-run expected average cost criterion and investigate the uniqueness and stability of minimizing Markov actions. More precisely, we prove that in a set of MDPs equipped with a natural complete metric subset of MDPs with essentially unique and stable minimizing Markov actions contains a countable intersection of open everywhere dense sets. Thus, the property of having essentially unique and stable minimizing Markov actions is generic for this class of MDPs.

The celebrated well-posedness notion for a model consists of three conditions: that the model will possess a solution, that this solution will be unique and that it will change continuously with respect to changes in the parameters that define it. This study is directly related to these three properties: we identify the class of models for which the first property (namely existence) holds, and for this class we study the question of uniqueness and stability of the optimal solutions. Thus Chap. 2 deals with very basic questions related to the study and definition of MDPs.

In order to study the well-posedness of our model we use the so-called generic approach which is applied fruitfully in many areas of Analysis (see, for example, [9, 83, 93, 95, 98] and the references mentioned there).

According to the generic approach we say that a property holds for a generic (typical) element of a complete metric space (or the property holds generically) if the set of all elements of the metric space possessing this property contains a G_δ everywhere dense subset of the metric space which is a countable intersection of open everywhere dense sets [58].

Note that the prototype of the main result of Chap. 2 was obtained in [68] under the assumption that the cost functions are convex. Here this assumption is not used.

A subclass of (MDPs), which was considered in Chap. 2, is studied in Chap. 3. It is shown that the asymptotic turnpike property for good programs on infinite horizon and the existence of an overtaking optimal policy follow from the uniqueness of minimizing Markov actions.

Chapter 4 is devoted to the analysis of the class of (MDPs) considered in Chap. 3. We show that the uniqueness of minimizing Markov actions implies the weak turnpike property for approximate optimal programs on finite horizon.

In Chap. 5 we consider the subclass of (MDPs) studied in Chap. 2 with convex action sets and convex cost functions. It is proved that if cost functions are strictly convex, then minimizing Markov actions are unique. It is also shown that a typical (in the sense of Baire category) cost function is in fact strictly convex.

Turnpike properties for MDPs with perturbations are studied in Chap. 6. Assuming the uniqueness of minimizing Markov actions we establish that the asymptotic turnpike property and the weak turnpike property are stable under the perturbations of cost functions. The prototype of the result on weak turnpike property is Theorem 1.9. Our next goal will be to obtain a result on a strong turnpike property for which Theorem 1.8 is a prototype.

It should be mentioned that in Chaps. 2–6 the study of our stochastic optimal control problem is reduced to the study of the corresponding deterministic optimal control problem for which A is the set of states, $\mathcal{M}$ is the control set and $r \in C(G)$ is a cost function. If at time $k \geq 0$ a state of the system is $x_k \in A$, then we choose $M^{(k)} \in \mathcal{M}$ and get

$$x^{(k+1)} = ((M^{(k)})^T x^{(k)}).$$

For a given natural number N the cost over the $[0, N]$ time interval is

$$\sum_{k=0}^{N-1} \sum_{j=1}^{m} x_j^{(k)} r(j, M_j^{(k)}).$$

This deterministic optimal control system is a particular case of a general optimal control system which will be studied in Chap. 8. The strong turnpike properties for deterministic optimal control systems are well-known in the literature [98, 100, 102] but they are related to states of the system. But for our MDPs and the related deterministic optimal control system the turnpike phenomenon holds both for its states and its controls. Therefore the existing turnpike results cannot be applied to our situation and new turnpike results are needed. They are presented in Chap. 8. As usual our strong turnpike property will be deduced from the asymptotic turnpike property and some controllability conditions which are discussed in Chap. 7.

Chapter 7 is devoted to the study of controllability properties of discrete-time deterministic control systems. These properties are important in the turnpike theory. We introduce a notion of an $\mathcal{M}$-positive matrix and show that under certain condition for a typical objective function the corresponding MDP has the unique minimizing Markov actions which form a $\mathcal{M}$-positive matrix. In this case our controllability results are true and the strong turnpike property holds and is stable. This is shown in Chap. 8.

Our results are obtained for a certain class of MDPs and now we present an example of MDP belonging to it. Actually we need to provide an example of a function r.

Example 1.13 Assume that for each $i \in \{1, \ldots, m\}$, $\theta(i, \cdot) : A(i) \to [0, \infty)$, $\widehat{a}^i \in A(i)$ is its unique minimizer satisfying

$$\theta(i, \widehat{a}^i) = 0.$$

Let $\lambda \in R^1$ and $I \in R^n$. For all $i \in \{1, \ldots, m\}$ and all $a \in A(i)$ set

$$r(i, a) = \theta(i, a) - a \cdot I + I_i + \lambda.$$

It is not difficult to see that for each $i \in \{1, \ldots, m\}$,

$$I_i + \lambda = \min_{a \in A(i)} [r(i, a) + a \cdot I] = r(i, \widehat{a}) + \widehat{a} \cdot I.$$

Example 1.14 Assume that for each $i \in \{1, \ldots, m\}$, $\widehat{a}^i \in A(i)$ and for each $a \in A(i)$,

$$\theta(i, a) = |\widehat{a}^i - a|_\infty.$$

For all $i \in \{1, \ldots, m\}$ and all $a \in A(i)$ set

$$r(i, a) = |\widehat{a}^i - a|_\infty - \sum_{k=1}^{m} a_k + 1.$$

It is not difficult to see that for each $i \in \{1, \ldots, m\}$,

$$1 = \min_{a \in A(i)} [r(i, a) + \sum_{k=1}^{m} a_k] = r(i, \widehat{a}) + \sum_{k=1}^{m} \widehat{a}_k.$$

Note that for each $i \in \{1, \ldots, m\}$ the set $A(i)$ is compact. Clearly, our results can be applied in the case when all these sets are finite.

All the main results of the book are new and represent at the first time in the literature the turnpike theory for the class of MDP's which is considered.

Chapter 2
Uniqueness and Stability of Optimal Policies of Finite State Markov Decision Processes

In this chapter we study infinite horizon discrete-time optimal control of Markov Decision Processes (MDPs) with finite state spaces and compact action sets and employ the long-run expected average cost criterion. We study the uniqueness and stability of minimizing Markov actions and show that in a set of MDPs equipped with a natural complete metric subset of MDPs with essentially unique and stable minimizing Markov actions contains a countable intersection of open everywhere dense sets. Thus, the property of having essentially unique and stable minimizing Markov actions is generic for this class of MDPs.

2.1 Preliminaries

Assume that $r \in C(G)$ and there exists $\lambda \in R^1$, $I = (I_1, \ldots, I_m) \in R^n$ such that

$$I_i + \lambda = \min_{a \in A(i)} [r(i, a) + a \cdot I], \ i = 1, \ldots m. \tag{2.1}$$

By (2.1) for each $i \in \{1, \ldots, m\}$ there exists $\widehat{a}^i \in A(i)$ such that

$$I_i + \lambda = r(i, \widehat{a}^i) + \widehat{a}^i \cdot I. \tag{2.2}$$

Equation (2.1) is well known in the optimal stochastic control [48] where it is used in order to show the existence of average cost optimality policies. Here this equation is used to order to establish the existence of overtaking optimal policies. The notion of overtaking optimality is essentially stronger than the notion of the average cost optimality.

For all $i \in \{1, \ldots, m\}$ and all $a \in A(i)$ set

$$\theta(i, a) = r(i, a) + a \cdot I - I_i - \lambda. \tag{2.3}$$

© The Author(s), under exclusive license to Springer Nature Switzerland AG 2025
A. J. Zaslavski, *Turnpike Phenomenon for Markov Decision Processes*,
SpringerBriefs in Mathematics, https://doi.org/10.1007/978-3-032-00854-1_2

By (2.1)–(2.3),

$$\theta(i, a) \geq 0, \ i \in \{1, \ldots, m\}, \ a \in A(i), \tag{2.4}$$

$$\theta(i, \widehat{a}^i) = 0, \ i \in \{1, \ldots, m\}.$$

Consider the matrix $\widehat{M} \in \mathcal{M}$ such that

$$\widehat{M}_i = \widehat{a}_i, \ i = 1, \ldots, m \tag{2.5}$$

and associate with $\widehat{M}$ the nonrandomized stationary Markovian policy $\widehat{\sigma} = \{\widehat{M}\}^{\infty}$.

Proposition 2.1 ([68], Proposition 2.1) *Let $x^{(0)} \in A$ and let $\sigma = \{M^{(k)}\}_{k=0}^{\infty}$ be any Markov policy. Then*

$$|C_N^r(x^{(0)}, \widehat{\sigma}) - \lambda N| \leq 2|I|_{\infty}$$

and

$$C_N^r(x^{(0)}, \sigma) - \lambda N \geq -2|I|_{\infty}$$

hold for every integer $N \geq 1$.

Proof Let $N \geq 1$ be an integer and for each integer $k \geq 0$,

$$x^{(k+1)} = (M^{(k)})^T x^{(k)}.$$

By (1.2) and (2.3),

$$C_N^r(x^{(0)}, \sigma) = \sum_{k=0}^{N-1} \sum_{j=1}^{m} x_j^{(k)} r(j, M_j^{(k)})$$

$$= \sum_{k=0}^{N-1} \sum_{j=1}^{m} x_j^{(k)} (\theta(j, M_j^{(k)}) - M_j^{(k)} \cdot I + I_l + \lambda)$$

$$= \lambda N + \sum_{k=0}^{N-1} \sum_{j=1}^{m} x_j^{(k)} \theta(j, M_j^{(k)}) + \sum_{k=0}^{N-1} (x^{(k)} \cdot I - \sum_{j=1}^{m} x_j^{(k)} \sum_{i=1}^{m} M_{j,i}^{(k)})$$

$$= \lambda N + \sum_{k=0}^{N-1} \sum_{j=1}^{m} x_j^{(k)} \theta(j, M_j^{(k)}) + \sum_{k=0}^{N-1} (x^{(k)} \cdot I - (M^{(k)})^T \cdot I)$$

$$= \lambda N + \sum_{k=0}^{N-1} \sum_{j=1}^{m} x_j^{(k)} \theta(j, M_j^{(k)}) + \sum_{k=0}^{N-1} (x^{(k)} \cdot I - x^{(k+1)} \cdot I)$$

$$= \lambda N + \sum_{k=0}^{N-1} \sum_{j=1}^{m} x_j^{(k)} \theta(j, M_j^{(k)}) + x^{(0)} \cdot I - x^{(N)} \cdot I.$$

This completes the proof of Proposition 2.1.

Corollary 2.2 *Let $x^{(0)} \in A$ be fixed. Then*

$$\lambda = \inf\{\liminf_{N \to \infty} N^{-1} C_N^r(x^{(0)}, \sigma) :$$

$$\sigma = \{M^{(k)}\}_{k=0}^{\infty} \text{ is a nonrandomized Markov policy}\}.$$

Denote by $C_{reg}(G)$ the set of all $r \in C(G)$ for which there exist $\lambda \in R^1$ and $I \in R^m$ satisfying (2.1).

If $r \in C_{reg}(G)$ and (I, λ) satisfies (2.1), then we set $\lambda(r) = \lambda$. Corollary 2.2 implies that $\lambda(r)$ is well defined and the mapping $r \to \lambda(r)$ on $C_{reg}(G)$ is continuous.

Let $i, j \in \{1, \ldots, m\}$. We say that i reaches j within one step if there exists $a \in A(i)$ such that $a_j > 0$.

Let $i, j \in \{1, \ldots m\}$ and $k \geq 2$ be an integer. We say that i reaches j within k steps if there is an $l \in \{1, \ldots, m\}$ such that i reaches l within $k - 1$ steps and l reaches j within one step.

Let $i, j \in \{1, \ldots m\}$. We say that i communicates with j if there exist integer $k \geq 1$ such that i reaches j within k steps.

Corollary 2.2 implies the following result.

Proposition 2.3 *Let $\Omega \subset C_{reg}(G)$, $\Omega \neq \emptyset$ and*

$$\sup\{\|r\| : r \in \Omega\} < \infty.$$

Then

$$\sup\{|\lambda(r)| : r \in \Omega\} < \infty.$$

For each $r \in C_{reg}(G)$ set

$$I(r) = \{I \in R^m : \quad (2.1) \text{ holds with } \lambda = \lambda(r)$$

$$\text{and } \max\{I_j : j \in \{1, \ldots, m\}\} = 0\}.$$

Clearly $I(r) \neq \emptyset$ for any $r \in C_{reg}(G)$.

Proposition 2.4 *Assume that i communicates with j for each pair of integers $i, j \in \{1, \ldots, m\}$. Let Ω be a nonempty, bounded subset of $C_{reg}(G)$, and let $M_0 > 0$. Then there exists a constant $K > 0$ such that for each $r \in \Omega$ and each $I \in R^m$ which satisfies*

$$\max\{I_j : j = 1, \ldots, m\} = 0 \tag{2.6}$$

and

$$I_i \leq M_0 + \min_{a \in A(i)} \{r(i, a) + a \cdot I\} \text{ for every } i \in \{1, \ldots, m\}, \tag{2.7}$$

the following relation holds:

$$|I_i| \leq K \text{ for every } i \in \{1, \ldots, m\}.$$

Proof For each $i, j \in \{1, \ldots, m\}$ such that i reaches j over one step there is $a^{(i)} \in A(i)$ such that $a_j^{(i)} > 0$. Let F be the collection of all these positive numbers $a_j^{(i)}$ and let $d_0 = \inf(F)$. We also denote

$$D_0 = \sup\{\|r\| : r \in \Omega\}$$

and choose

$$K > (1 + M_0 + D_0) \sum_{l=1}^{m} d_0^{-l}.$$

Assume that $r \in \Omega$ and $I \in R^m$ satisfies (2.6) and (2.7). Then there exists $i_0 \in \{1, \ldots, m\}$ such that $I_{i_0} = 0$. The validity of the proposition follows from the following assertion which is proved by induction:

For each integer $k = 1, \ldots, m - 1$, if $j \in \{1, \ldots, m\}$, $i_0 \neq j$, i_0 reaches j over k steps, then

$$I_j \geq (-M_0 - D_0)(d_0^{-1} + \cdots + d_0^{-k}).$$

Let $k = 1$, $j \in \{1, \ldots, m\}$, $i_0 \neq j$, i_0 reaches j over one step. There exists $a^{(i_0)} \in A(i_0)$ such that $a_j^{(i_0)} > 0$. By (2.6) and (2.7),

$$0 = I_{i_0} \leq M_0 + r(i_0, a^{(i_0)}) + a^{(i_0)} \cdot I$$

$$\leq M_0 + D_0 + a_j^{(i_0)} I_j$$

and

$$I_j \geq (-M_0 - D_0)(a_j^{(i_0)})^{-1} \geq (-M_0 - D_0)d_0^{-1}$$

and our assertion holds for $k = 1$.

Assume that our assertion holds for $k \in \{1, \ldots, m\}$ where $k < m - 1$, $l \in \{1, \ldots, m\}$ and that i_0 reaches l over $k + 1$ steps. Then there exists $j \in \{1, \ldots, m\}$ such that i_0 reaches j over k steps and j reacher l over one step. Then since our assertion holds for k we have

$$I_j \geq (-M_0 - D_0)(d_0^{-1} + \cdots + d_0^{-k}).$$

There exists $a^{(j)} \in A(j)$ such that

$$a_l^{(j)} > 0.$$

By (2.6), (2.7) and the relations above,

$$(-M_0 - D_0)(d_0^{-1} + \cdots + d_0^{-k})$$

$$\leq M_0 + r(j, a^{(j)}) + a^{(j)} \cdot I$$

$$\leq M_0 + D_0 + a_l^{(j)} I_l$$

and

$$I_l \geq (-M_0 - D_0)(1 + d_0^{-1} + \cdots + d_0^{-k})(a_l^{(j)})^{-1} \geq (-M_0 - D_0)(d_0^{-1} + \cdots + d_0^{-k-1}).$$

Thus our assertion holds for $k + 1$ too. Therefore it holds for all $k = 1, \ldots, m - 1$. Proposition 2.4 is proved.

Theorem 2.5 *Assume that all $i, j \in \{1, \ldots, m\}$ communicate. Then the set $C_{reg}(G)$ is a closed subset of $C(G)$.*

Proof Assume that $\{r_k\}_{k=1}^{\infty} \subset C_{reg}$, $r \in C(G)$ and that

$$\lim_{k \to \infty} \|r_k - r\| = 0.$$

In view of Corollary 2.2 there exists

$$\bar{\lambda} = \lim_{k \to \infty} \lambda(r_k).$$

For each integer $k \geq 1$, there exists $I^{(k)} \in R^m$ such that

$$I_i^{(k)} + \lambda(r_k) = \min_{a \in A(i)} [r_k(i, a) + a \cdot I^{(k)}], \quad i = 1, \ldots m,$$

$$\max\{I_j^{(k)} : j = 1, \ldots, m\} = 0$$

and for each $i \in \{1, \ldots, m\}$ there exists $a^{(k,i)} \in A(i)$ such that

$$I_i^{(k)} + \lambda(r_k) = r_k(i, a^{(k,i)}) + a^{(k,i)} \cdot I^{(k)}.$$

Proposition 2.4 implies that the sequence $\{I^{(k)}\}_{k=1}^{\infty}$ is bounded. Extracting a subsequence and re-indexing if necessary, we may assume without loss of generality that

$$I^{(k)} \to \bar{I}, \ a^{(k,i)} \to \bar{a}^{(i)} \in A(i), \ i = 1, \ldots, m$$

as $k \to \infty$. The relations above imply that for each $i \in \{1, \ldots, m\}$,

$$\bar{I}_i + \bar{\lambda} = r(i, \bar{a}^{(i)}) + \bar{a}^{(i)} \cdot \bar{I},$$

$$\bar{I}_i + \bar{\lambda} \leq \min_{a \in A(i)} [r(i, a) + a \cdot \bar{I}], \ i = 1, \ldots m.$$

Therefore $r \in C_{reg}(G)$ and Theorem 2.5 is proved.

2.2 A Generic Result

In this section we continued to use the notation and definitions introduced in Sects. 2.1 and 2.2.

We consider the topological space $\mathcal{M}$ equipped with the topology induced by the Euclidean metric of $R^{m \times m}$.

Assume that all $i, j \in \{1, \ldots, m\}$ communicate. Then by Theorem 2.5, the set $C_{reg}(G)$ is a closed subset of $C(G)$ and we consider the topological space $C_{reg}(G) \subset C(G)$ with the relative topology.

A matrix $M \in \mathcal{M}$ is called good if for each pair $i, j \in \{1, \ldots, m\}$ there exists a natural number k and a sequence $i_0, \ldots, i_k \in \{1, \ldots, m\}$ such that $i_0 = i, i_k = j$ and

$$M_{i_s i_{s+1}} > 0, \ s = 0, \ldots, k - 1.$$

We use the following condition.

Condition C. The set of all good matrices is an everywhere dense subset of $\mathcal{M}$ equipped with the natural topology (induced by the Euclidean metric).

Note that if all the sets $A(i), i = 1, \ldots, m$ are finite, then the set $\mathcal{M}$ is finite too and in this case condition C holds if every matrix in $\mathcal{M}$ is good.

Condition C plays an important role in our study. Now we show that it holds if all the sets $A(i), i = 1, \ldots, m$ are convex.

Proposition 2.6 *Assume that for each $i \in \{1, \ldots, m\}$ the set $A(i)$ is convex. Then the condition C holds.*

Proof It is easy to see that if $\alpha \in (0, 1)$, $M^{(1)}$, $M^{(2)} \in \mathcal{M}$ and $M^{(1)}$ is good, then $\alpha M^{(1)} + (1 - \alpha) M^{(2)}$ belongs to $\mathcal{M}$ and it is good. Therefore in order to prove the proposition it is sufficient to show the existence of a good matrix in $\mathcal{M}$.

Let $i \in \{1, \ldots, m\}$. Since the set $A(i)$ is convex there is $\widehat{a}^i \in A(i)$ such that

$$\{j \in \{1, \ldots, m\} : \ b_j > 0\} \subset \{j \in \{1, \ldots, m\} : \ \widehat{a}^i_j > 0\}$$

for all $b = (b_1, \ldots, b_m) \in A(i)$. Consider the matrix $\widehat{M} \in \mathcal{M}$ with $\widehat{M}_i = \widehat{a}^i$, $i = 1, \ldots, m$. Since all $i, j \in \{1, \ldots, m\}$ communicate it follows from the choice of $\widehat{a}^i$, $i = 1, \ldots, m$ that $\widehat{M}$ is good. Proposition 2.6 is proved.

If for all $i = 1, \ldots, m$ the set $A(i)$ is convex, then we denote by $C_{conv}(G)$ the set of all $r \in C_{reg}(G)$ such that $r(i, \cdot) : A(i) \to R^1$ is convex for all $i = 1, \ldots, m$. In this case $C_{conv}(G)$ is a closed subset of $C_{reg}(G)$ and we consider the topological subspace $C_{conv}(G) \subset C_{reg}(G)$ equipped with the relative topology.

From now on we assume that the Condition C holds. The following theorem is the main result of this chapter. It shows that for a generic cost function equation (2.1) has a unique solution.

Theorem 2.7 *Let $\mathcal{A}$ be either $C_{reg}(G)$ or $C_{conv}(G)$ (if $A(i)$ is convex for all $i = 1, \ldots, m$). Then there exists a set $\mathcal{F} \subset \mathcal{A}$ which is a countable intersection of open everywhere dense subsets of $\mathcal{A}$ such that for each $r \in \mathcal{F}$ the following property holds:*

There exist actions $a^{(r,i)} \in A(i)$, $i = 1, \ldots, m$ such that the matrix $M \in \mathcal{M}$ with $M_i = a^{(r,i)}$, $i = 1, \ldots, m$ is good. If $I \in R^m$, $a^i \in A(i)$, $i = 1, \ldots, m$ satisfy

$$I_i + \lambda(r) = \min_{a \in A(i)} [r(i, a) + a \cdot I] = r(i, a^i) + a^i \cdot I, \ i = 1, \ldots, m,$$

then $a^i = a^{(r,i)}, i = 1, \ldots, m$.

2.3   Auxiliary Results for Theorem 2.7

Lemma 2.8 *Assume that $r \in C_{reg}(G)$ and $\epsilon > 0$. Then there exist $\tilde{r} \in C_{reg}(G)$, $\bar{I} \in R^m$, $b^i \in A(i)$, $i = 1, \ldots, m$ and $M \in \mathcal{M}$ such that*

$$\|\tilde{r} - r\| \leq \epsilon, \tag{2.8}$$

$$\bar{I}_i + \lambda(\tilde{r}) = \min_{a \in A(i)} [\tilde{r}(i, a) + a \cdot \bar{I}] = \tilde{r}(i, b^i) + b^i \cdot \bar{I}, \ i = 1, \ldots, m, \tag{2.9}$$

$$M_i = b^i, \ i = 1, \ldots, m$$

and M is good. Moreover if for all $i \in \{1, \ldots, m\}$, $A(i)$ is convex and $r(i, \cdot) :$ $A(i) \to R^1$ is convex for all $i = 1, \ldots, m$, then $\tilde{r}(i, \cdot) : A(i) \to R^1$ is convex for all $i = 1, \ldots, m$.

Proof Since $r \in C_{reg}(G)$ there exist $\bar{I} \in R^m$, $\bar{a}^i \in A(i)$, $i = 1, \ldots, m$ such that

$$\bar{I}_i + \lambda(r) = \min_{a \in A(i)} [r(i, a) + a \cdot \bar{I}] = r(i, \bar{a}^i) + \bar{a}^i \cdot \bar{I}, \; i = 1, \ldots, m. \tag{2.10}$$

Define

$$\theta(i, a) = r(i, a) + a \cdot \bar{I} - \bar{I}_i - \lambda(r), \; i \in \{1, \ldots, m\}, \; a \in A(i). \tag{2.11}$$

By (2.10) and (2.11),

$$\theta(i, a) \geq 0, \; i \in \{1, \ldots, m\}, \; a \in A(i),$$

$$\theta(i, \bar{a}^i) = 0, \; i = 1, \ldots, m. \tag{2.12}$$

Clearly, $\theta : G \to R^1$ is continuous. Therefore there is $\delta \in (0, 4^{-1})$ such that

$$\theta(i, a) \leq 8^{-1}\epsilon \text{ for each } (i, a) \in G \text{ satisfying } \|a - \bar{a}^i\| \leq 2\delta. \tag{2.13}$$

In view of condition (C) there exists a good matrix $M \in \mathcal{M}$ such that

$$\|M_i - \bar{a}^i\| \leq \delta, \; i = 1, \ldots, m. \tag{2.14}$$

Set

$$b^i = M_i, \; i = 1, \ldots, m. \tag{2.15}$$

For $(i, a) \in G$ set

$$\tilde{\theta}(i, a) = \max\{\theta(i, a) - \theta(i, b^i), 0\}. \tag{2.16}$$

By (2.13)–(2.15), for $i = 1, \ldots, m$,

$$\theta(i, b^i) = \theta(i, M_i) \leq 8^{-1}\epsilon. \tag{2.17}$$

It follows from (2.16) and (2.17) that for all $(i, a) \in G$,

$$-8^{-1}\epsilon + \theta(i, a) \leq \tilde{\theta}(i, a) \leq \theta(i, a) \tag{2.18}$$

and

$$\tilde{\theta}(i, b^i) = 0, \; i = 1, \ldots, m. \tag{2.19}$$

Define for all $(i, a) \in G$

$$\tilde{r}(i, a) = \tilde{\theta}(i, a) - a \cdot \bar{I} + \bar{I}_i + \lambda(r).$$ (2.20)

It is easy to see that $\tilde{r} : G \to R^1$ is a continuous function and if for all $i = 1, \ldots, m$ $A(i)$ is a convex set and $r(i, \cdot) : A(i) \to R^1$ is a convex function, then $\tilde{r}(i) : A(i) \to R^1$ is a convex function for all $i = 1, \ldots, m$.

It follows from (2.11), (2.18) and (2.20) that for all $(i, a) \in G$

$$|\tilde{r}(i, a) - r(i, a)| = |\tilde{\theta}(i, a) - \theta(i, a)| \le \epsilon/8,$$

$\|\tilde{r} - r\| \le \epsilon$ and (2.8) holds.

By (2.16) and (2.20) for each $i = 1, \ldots, m$,

$$\tilde{r}(i, a) + a \cdot \bar{I} \ge \bar{I}_i + \lambda(r) \text{ for all } a \in A(i)$$

and

$$\bar{I}_i + \lambda(r) \le \min_{a \in A(i)} [\tilde{r}(i, a) + a \cdot \bar{I}].$$ (2.21)

In view of (2.19) and (2.20) for $i = 1, \ldots, m$,

$$\tilde{r}(i, b^i) + b^{(i)} \cdot \bar{I} = \bar{I}_i + \lambda(r).$$

Together with (2.21) this implies that for $i = 1, \ldots, m$

$$\bar{I}_i + \lambda(r) = \min_{a \in A(i)} [\tilde{r}(i, a) + a \cdot \bar{I}] = \tilde{r}(i, b^i) + b^i \cdot \bar{I}.$$

This implies that $\tilde{r} \in C_{reg}(G)$, $\lambda(\tilde{r}) = \lambda(r)$ and (2.9) holds. This completes the proof of Lemma 2.8.

Lemma 2.9 *Let* $r \in C_{reg}(G)$, $\gamma \in (0, 1)$, $I \in R^m$, $\bar{a}^i \in A(i)$, $i = 1, \ldots, m$ *be such that*

$$I_i + \lambda(r) = \min_{a \in A(i)} \{r(i, a) + a \cdot I\} = r(i, \bar{a}^i) + \bar{a}^i \cdot I, \ i = 1, \ldots, m$$ (2.22)

and let $\bar{M} \in \mathcal{M}$ *be such that* $\bar{M}_i = \bar{a}^i$, $i = 1, \ldots, m$. *Assume the matrix* $\bar{M}$ *is good and define*

$$r_\gamma(i, a) = r(i, a) + \gamma \|a - \bar{a}^i\|, \ (i, a) \in G.$$ (2.23)

Then

$$r_\gamma \in \tilde{C}_{reg}(G), \ \lambda(r_\gamma) = \lambda(r)$$ (2.24)

and for $i = 1, \ldots, m$

$$I_i + \lambda(r) = \min_{a \in A(i)} [r_\gamma(i, a) + a \cdot I] = r_\gamma(i, \bar{a}^i) + \bar{a}^i \cdot I \tag{2.25}$$

and for $i = 1, \ldots, m$ if $A(i)$ is a convex set and $r(i, \cdot) : A(i) \to R^1$ is a convex function, then $r_\gamma(i, \cdot)$ is a convex function.

Moreover, if $\bar{I} \in R^m$, $b^i \in A(i)$, $i = 1, \ldots, m$ satisfy for $i = 1, \ldots, m$

$$\bar{I}_i + \lambda(r) = \min_{a \in A(i)} \{r_\gamma(i, a) + a \cdot \bar{I}\} = r_\gamma(i, b^i) + b^i \cdot \bar{I}, \tag{2.26}$$

then $b^i = \bar{a}^i$ for $i = 1, \ldots, m$.

Proof Clearly (2.24) and (2.25) hold. Assume that $\bar{I} \in R^m$, $b^i \in A(i)$, $i = 1, \ldots, m$ satisfy (2.26) for $i = 1, \ldots, m$. In order to complete the proof it is sufficient to show that $b^i = \bar{a}^i$, $i = 1, \ldots, m$.

Denote by M the $m \times m$ matrix such that $M_i = b^{(i)}$, $i = 1, \ldots, m$. Let $\sigma = \{M^\infty\}$ be a stationary Markov policy, $x^{(0)} = m^{-1}(1, 1, \ldots, 1)$ and let $\{\xi^{(k)}\}_{k=0}^\infty$ be the Markov chain associated with $(x^{(0)}, \sigma)$.

For $k = 0, 1, \ldots$ we denote by $x^{(k)}$ the distribution of $\xi^{(k)}$ and thus we have relation $x^{(k+1)} = M^T x^{(k)}$ where M^T is a transpose of M.

It follows from (2.24), (2.26) and Proposition 2.1 that there is $c_0 > 0$ such that for all integers $N \geq 1$

$$|C_N^{r\gamma}(x^{(0)}, \sigma) - \lambda(r)N| \leq c_0,$$

$$C_N^r(x^{(0)}, \sigma) - \lambda(r)N \geq -c_0. \tag{2.27}$$

By (1.2), (2.23) and (2.27) for all natural numbers N,

$$c_0 \geq C_N^{r\gamma}(x^{(0)}, \sigma) - \lambda(r)N = \sum_{k=0}^{N-1}\sum_{j=1}^{m} x_j^{(k)} r_\gamma(j, b^j) - \lambda(r)N$$

$$= \sum_{k=0}^{N-1}\sum_{j=1}^{m} x_j^{(k)} r(j, b^j) - \lambda(r)N + \sum_{k=0}^{N-1}\sum_{j=1}^{m} x_j^{(k)} \gamma \|\bar{a}^j - b^j\|$$

$$= C_N^r(x^{(0}, \sigma) - \lambda(r)N + \sum_{k=0}^{N-1}\sum_{j=1}^{m} x_j^{(k)} \gamma \|\bar{a}^j - b^j\|$$

$$\geq -c_0 + \sum_{k=0}^{N-1}\sum_{j=1}^{m} x_j^{(k)} \gamma \|\bar{a}^j - b^j\|$$

and

$$\sum_{k=0}^{N-1}\sum_{j=1}^{m} x_j^{(k)}\|\bar{a}^j - b^j\| \le 2c_0/\gamma \text{ for all natural numbers } N. \tag{2.28}$$

Denote by S the set of all $j \in \{1, \ldots, m\}$ for which

$$\limsup_{k\to\infty} x_j^{(k)} > 0. \tag{2.29}$$

Clearly $S \ne \emptyset$. Relations (2.28) and (2.29) imply that for all $j \in S$,

$$b^j = \bar{a}^j. \tag{2.30}$$

By (2.29), (2.30), equality $x^{(k+1)} = M^T x^{(k)}$, $k = 0, 1, \ldots$ the following property holds:

(P1)   If $j \in S, i \in \{1, \ldots, m\}$, $\bar{a}_i^j > 0$. then $i \in S$.
Since $\bar{M}$ is good and $S \ne \emptyset$, property (P1) implies that $S = X$. Together with (2.30) this implies that $b^j = \bar{a}^j$, $j = 1, \ldots, m$. Lemma 2.9 is proved.

It is easy to see that the following lemma holds.

Lemma 2.10 *Let $\tilde{M} \in \mathcal{M}$ be a good matrix. Then there is $\epsilon > 0$ such that each $M \in \mathcal{M}$ satisfying $\|\tilde{M}_i - M_i\| \le \epsilon$, $i = 1, \ldots, m$ is also good.*

Proposition 2.11 *Assume that $\bar{r} \in C_{reg}(G)$ and there exist $\bar{a}^i \in A(i)$, $i = 1, \ldots, m$ such that the following property holds:*
If $\bar{I} = (\bar{I}_1, \ldots, \bar{I}_m) \in R^m$, $a^i \in A(i)$, $i = 1, \ldots, m$ satisfy

$$\bar{I}_i + \lambda(\bar{r}) = \min_{a\in A(i)} [\bar{r}(i, a) + a \cdot \bar{I}] = \bar{r}(i, a^i) + a^i \cdot \bar{I}, \; i = 1, \ldots, m,$$

then $a^i = \bar{a}^i$, $i = 1, \ldots, m$.
Let $\epsilon > 0$. Then there exists $\delta > 0$ such that for each $r \in C_{reg}(G)$ which satisfies $\|r - \bar{r}\| \le \delta$, each $I \in R^m$ and each $\{b^i\}_{i=1}^m \subset A$ which satisfy

$$b^i \in A(i), \; i = 1, \ldots, m,$$

$$I_i + \lambda(r) = \min_{a\in A(i)} [r(i, a) + a \cdot I] = r(i, b^i) + b^i \cdot I, \; i = 1, \ldots, m$$

the inequalities $\|b^i - \bar{a}^{(i)}\| \le \epsilon$, $i = 1, \ldots, m$ hold.

Proof Assume that the proposition does not hold. Then there exist sequences $\{r_n\}_{n=1}^\infty \subset C_{reg}(G)$, $I^{(n)} \in R^m$, $n = 1, 2, \ldots$ and for each integer $n \ge 1$ there exist actions $\{b^{(n,i)}\}_{i=1}^{m} \subset A$, $n = 1, 2, \ldots$ such that

$$\lim_{n \to \infty} \|r_n - \bar{r}\| = 0, \tag{2.31}$$

$$I_i^{(n)} + \lambda(r_n) = \min_{a \in A(i)} [r_n(i, a) + a \cdot I^{(n)}], \ i = 1, \ldots, m, \ n = 1, 2, \ldots, \tag{2.32}$$

$$I_i^{(n)} + \lambda(r_n) = r_n(i, b^{n,i}) + b^{(n,i)} \cdot I^{(n)}, \ b^{(n,i)} \in A(i), \tag{2.33}$$

$n = 1, 2, \ldots, \ i = 1, \ldots, m,$

$$\max\{I_i^{(n)} : \ i = 1, \ldots, m\} = 0, \ n = 1, 2, \ldots \tag{2.34}$$

and

$$\max\{\|\bar{a}^{(i)} - b^{n,i}\| : \ i = 1, \ldots, m\} \geq \epsilon, \ n = 1, 2, \ldots. \tag{2.35}$$

Proposition 2.4 and (2.31), (2.32) imply that the sequence $\{I^{(n)}\}_{n=1}^{\infty} \subset R^n$ is bounded. We may assume without loss of generality that there exist

$$\bar{b}^{(i)} = \lim_{n \to \infty} b^{(n,i)}, \ i = 1, \ldots, m, \ \bar{I} = \lim_{n \to \infty} I^{(n)}. \tag{2.36}$$

It follows from (2.31)–(2.33) and (2.36) that $\bar{b}^{(i)} \in A(i)$, $i = 1, \ldots, m$,

$$\bar{I}_i + \lambda(\bar{r}) = \min_{a \in A(i)} [\bar{r}(i, a) + a \cdot \bar{I}], \ i = 1, \ldots, m,$$

$$\bar{I}_i + \lambda(\bar{r}) = \bar{r}(i, \bar{b}^{(i)}) + \bar{b}^{(i)} \cdot \bar{I}, \ i = 1, \ldots, m.$$

By (2.35) and (2.36),

$$\max\{\|\bar{a}^{(i)} - \bar{b}^{(i)}\| : \ i = 1, \ldots, m\} \geq \epsilon.$$

On the other hand by our assumptions and the relations above,

$$\bar{b}^{(i)} = \bar{a}^{(i)}, \ i = 1, \ldots, m.$$

The contradiction we have reached proves Proposition 2.11.

2.4   Proof of Theorem 2.7

Let $r \in \mathcal{A}$, $\gamma \in (0, 1)$. By Lemma 2.8 there exist $r_{\gamma,0} \in \mathcal{A}$, $I \in R^m$, $a^{(r,i)} \in A(i)$, $i = 1, \ldots, m$ such that

$$\|r_{\gamma,0} - r\| \leq \gamma/2, \tag{2.37}$$

$$I_i + \lambda(r_{\gamma,0}) = \min_{a \in A(i)} [r_{\gamma,0}(i, a) + a \cdot I] = r_{\gamma,0}(i, a^{(r,i)}) + a^{(r,i)} \cdot I, \; i = 1, \ldots, m, \tag{2.38}$$

the matrix M with

$$M_i = a^{(r,i)}, \; i = 1, \ldots, m \tag{2.39}$$

is good. Define

$$r_\gamma(i, a) = r_{\gamma,0}(i, a) + \gamma \|a - a^{(r,i)}\|, \; (i, a) \in G. \tag{2.40}$$

By (2.38)–(2.40), Lemma 2.9 and since M is good

$$r_\gamma \in \mathcal{A},$$

$$I_i + \lambda(r_\gamma) = \min_{a \in A(i)} [r_\gamma(i, a) + a \cdot I] = r_\gamma(i, a^{(r,i)}) + a^{(r,i)} \cdot I, \; i = 1, \ldots, m \tag{2.41}$$

and the following property holds:

(P2)   if

$$\tilde{I} = (\tilde{I}_1, \ldots, \tilde{I}_m) \in R^m, \; b^i \in A(i), \; i = 1, \ldots, m,$$

$$\tilde{I}_i + \lambda(r_\gamma) = \min_{a \in A(i)} [r_\gamma(i, a) + a \cdot \tilde{I}] = r_\gamma(i, b^i) + b^i \cdot \tilde{I}, \; i = 1, \ldots, m,$$

then $b^i = a^{(r,i)}$, $i = 1, \ldots, m$.

Let $p \geq 1$ be an integer. In view of (P2), Proposition 2.11, Lemma 2.10 and since M is good there exists an open neighborhood $\mathcal{U}(r, \gamma, p)$ of r_γ in $\mathcal{A}$ such that the following properties hold:

(P3)   If

$$h \in \mathcal{U}(r, \gamma, p), \; I \in R^m, \; b^i \in A(i), \; i = 1, \ldots, m,$$

$$I_i + \lambda(h) = \min_{a \in A(i)} [h(i, a) + a \cdot I] = h(i, b^i) + b^i \cdot I, \; i = 1, \ldots, m,$$

then

$$\|b^i - a^{(r,i)}\| \le (4p)^{-1}, \ i = 1, \ldots, m;$$

(P4)   the matrix D with $D_i = b^i, i = 1, \ldots, m$, is good.

Define

$$\mathcal{F} = \cap_{k=1}^{\infty} \cup \{\mathcal{U}(r, \gamma, k) : \ r \in \mathcal{A}, \ \gamma \in (0, 1)\}. \tag{2.42}$$

Clearly, $\mathcal{F}$ is a countable intersection of open everywhere dense subsets of $\mathcal{A}$.
Assume that

$$h \in \mathcal{F}, \ I^{(q)} \in R^m, \ q = 1, 2, \ b^{(q,i)} \in A(i), \ i = 1, \ldots, m, \ q = 1, 2, \tag{2.43}$$

$$I_i^{(q)} + \lambda(h) = \min_{a \in A(i)} [h(i, a) + a \cdot I^{(q)}] = h(i, b^{(q,i)}) \cdot I^{(q)},$$

$$i = 1, \ldots, m, \ q = 1, 2. \tag{2.44}$$

For $q = 1, 2$ let $M^{(q)} \in \mathcal{M}$ with $M_i^{(q)} = b^{(q,i)}, i = 1, \ldots, m$. By (2.42)–(2.44) and property (P4), $M^{(1)}, M^{(2)}$ are good.

In order to complete the proof we need only to show that $b^{(1,i)} = b^{(2,i)}, i = 1, \ldots, m$. Let $k \ge 1$ be an integer. By (2.42)–(2.44), there exist $r_k \in \mathcal{A}, \gamma_k \in (0, 1)$ such that

$$h \in \mathcal{U}(r_k, \gamma_k, k).$$

It follows from this inclusion, (2.43), (2.44) and (P3) that for $i = 1, \ldots, m, q = 1, 2$

$$\|b^{(q,i)} - a^{(r_k,i)}\| \le (4k)^{-1}, \ i = 1, \ldots, m.$$

This implies that

$$\|b^{(1,i)} - b^{2,i)}\| \le (2k)^{-1}, \ i = 1, \ldots, m.$$

Since k is an arbitrary natural number we conclude that

$$b^{(1,i)} = b^{(2,i)}, \ i = 1, \ldots, m.$$

Theorem 2.7 is proved.

Chapter 3
Existence of an Overtaking Optimal Policy

In this chapter we consider a subclass of (MDPs) studied in Chap. 2 and assuming the uniqueness of minimizing Markov actions show the existence of an overtaking optimal policy. We also establish the asymptotic turnpike property for good programs on infinite horizon.

3.1 The Second Generic Result

In this chapter we use the notation and definitions introduced in Sects. 2.1–2.3 and use the assumptions introduced there.

Let $e = (1, \dots, 1) \in R^m$ be the vector all of its coordinates are equal to 1. We say that $x = (x_1, \dots, x_m) \in R^m$ satisfies $x >> 0$ if and only if $x_i > 0$, $i = 1, \dots, m$. Set

$$R^n_+ = \{x = (x_1, \dots, x_m) \in R^m : x_i \geq 0, \ i = 1, \dots, m\}.$$

For each matrix B of the order $m \times m$ set $B^1 = B$ and $B^{i+1} = B \times B^i$ for each integer $i \geq 1$.

Let $M \in \mathcal{M}$. Clearly, $Me = e$. It is well-known that there exists $z(M) \in R^m_+$ such that

$$z(M) \cdot e = 1, \ M^T z(M) = z(M). \tag{3.1}$$

Let $k \geq 1$ be an integer. A matrix $M \in \mathcal{M}$ is called k-positive if all the entries of M^k are positive.

The following result is well-known in the matrix theory [1].

A. J. Zaslavski, *Turnpike Phenomenon for Markov Decision Processes*,
SpringerBriefs in Mathematics, https://doi.org/10.1007/978-3-032-00854-1_3

Proposition 3.1 *Assume that k is a natural number and $M \in \mathcal{M}$ is k-positive. Then M is good, $z(M) >> 0$ and for each $i, j \in \{1, \ldots, m\}$,*

$$\lim_{p \to \infty} (M^T)^p_{i,j} = z_i(M). \tag{3.2}$$

Clearly, if $M \in \mathcal{M}$ is good, $z(M) >> 0$ and (3.2) holds for each $i, j \in \{1, \ldots, m\}$, then M is k-positive for some natural number k.

We use the following condition D.

Condition D. The set $\cup_{k=1}^{\infty} \{M \in \mathcal{M}$ is $k - $positive$\}$ is an everywhere dense subset of $\mathcal{M}$ equipped with the natural topology (induced by the Euclidean metric).

Clearly, condition D implies condition C. Note that if all the sets $A(i)$, $i = 1, \ldots, m$ are finite, then the set $\mathcal{M}$ is finite too and in this case condition D holds if every matrix in $\mathcal{M}$ is k-positive with some natural number k.

Let $M = (M_{i,j})$, $Q = (Q_{i,j})$, $i, j \in \{1, \ldots, m\}$ be $m \times m$ matrices. Set

$$|M|_{\infty} = \max\{|M_{i,j}| : \ i, j = 1, \ldots, m\}.$$

We say that $Q \leq M$ if and only if $M_{i,j} \leq Q_{i,j}$ for each $i, j \in \{1, \ldots, m\}$. For each $x = (x_1, \ldots, x_m)$, $y = (y_1, \ldots, y_m) \in R^m$ we say that $x \leq y$ if and only if $x_i \leq y_i$, $i = 1, \ldots, m$.

For each $z = (z_1, \ldots, z_m) \in R^m$ denote by $B(z) = (B(z)_{i,j})^m_{i,j=1}$ the $m \times m$ matrix such that for each $i, j \in \{1, \ldots, m\}$,

$$B(z)_{i,j} = z_i. \tag{3.3}$$

Proposition 3.2 *Assume that for each $i \in \{1, \ldots, m\}$ the set $A(i)$ is convex, k is a natural number, $\tilde{M} \in \mathcal{M}$ and all the entries of $\tilde{M}^k$ are positive. Then condition D holds.*

Proof It is easy to see that if $\alpha \in (0, 1)$ and $M \in \mathcal{M}$, then $M_\alpha := \alpha M + (1-\alpha)\tilde{M} \in \mathcal{M}$ and the entries of M_α^k are positive. Proposition 3.2 is proved.

In this chapter we prove the following improvement of Theorem 2.7. It shows that for a generic cost function equation (2.1) has a unique solution which forms a k-positive matrix.

Theorem 3.3 *Let $\mathcal{A}$ be either $C_{reg}(G)$ or $C_{conv}(G)$ (if $A(i)$ is convex for all $i = 1, \ldots, m$). Assume that condition (D) holds. Then there exists a set $\mathcal{F} \subset \mathcal{A}$ which is a countable intersection of open everywhere dense subsets of $\mathcal{A}$ such that for each $r \in \mathcal{F}$ the following property holds:*

There exist actions $a^{(r,i)} \in A(i)$, $i = 1, \ldots, m$ and a natural number k such that the matrix $M \in \mathcal{M}$ with $M_i = a^{(r,i)}$, $i = 1, \ldots, m$ is k-positive. If $I \in R^m$, $a^i \in A(i)$, $i = 1, \ldots, m$ satisfy

$$I_i + \lambda(r) = \min_{a \in A(i)} [r(i, a) + a \cdot I] = r(i, a^i) + a^i \cdot I, \ i = 1, \ldots, m,$$

then $a^i = a^{(r,i)}$, $i = 1, \ldots, m$.

In the sequel we use the next auxiliary result.

Lemma 3.4 *Assume that condition (D) holds, $r \in C_{reg}(G)$ and $\epsilon > 0$. Then there exist $\bar{r} \in C_{reg}(G)$, $\bar{I} \in R^m$, $b^i \in A(i)$, $i = 1, \ldots, m$ and $M \in \mathcal{M}$ such that*

$$\|\bar{r} - r\| \le \epsilon,$$

$$\bar{I}_i + \lambda(\bar{r}) = \min_{a \in A(i)} [\bar{r}(i, a) + a \cdot \bar{I}] = \bar{r}(i, b^i) + b^i \cdot \bar{I}, \ i = 1, \ldots, m,$$

$$M_i = b^i, \ i = 1, \ldots, m$$

and M is k-positive for some natural number k. Moreover if for all $i \in \{1, \ldots, m\}$, $A(i)$ is convex and $r(i, \cdot) : A(i) \to R^1$ is convex for all $i = 1, \ldots, m$, then $\bar{r}(i, \cdot) : A(i) \to R^1$ is convex for all $i = 1, \ldots, m$.

This result is proved exactly as Lemma 2.8.

Proof of Theorem 3.3 Let $r \in \mathcal{A}$, $\gamma \in (0, 1)$. By Lemma 3.4 there exist a natural number k, $r_{\gamma,0} \in \mathcal{A}$, $I \in R^m$, $a^{(r,i)} \in A(i)$, $i = 1, \ldots, m$ such that

$$\|r_{\gamma,0} - r\| \le \gamma/2, \tag{3.4}$$

$$I_i + \lambda(r_{\gamma,0}) = \min_{a \in A(i)} [r_{\gamma,0}(i, a) + a \cdot I] = r_{\gamma,0}(i, a^{(r,i)}) + a^{(r,i)} \cdot I, \ i = 1, \ldots, m,$$
$$\tag{3.5}$$

the matrix $\tilde{M}$ with

$$\tilde{M}_i = a^{(r,i)}, \ i = 1, \ldots, m \tag{3.6}$$

is k-positive. Define

$$r_\gamma(i, a) = r_{\gamma,0}(i, a) + \gamma \|a - a^{(r,i)}\|, \ (i, a) \in G. \tag{3.7}$$

By (3.5)–(3.7), Lemma 2.9 and since $\tilde{M}$ is good

$$r_\gamma \in \mathcal{A},$$

$$I_i + \lambda(r_\gamma) = \min_{a \in A(i)} [r_\gamma(i, a) + a \cdot I] = r_\gamma(i, a^{(r,i)}) + a^{(r,i)} \cdot I, \ i = 1, \ldots, m$$

and the following property holds:

(P1)   if

$$\tilde{I} = (\tilde{I}_1, \ldots, \tilde{I}_m) \in R^m, \ b^i \in A(i), \ i = 1, \ldots, m,$$

$$\tilde{I}_i + \lambda(r_\gamma) = \min_{a \in A(i)} [r_\gamma(i, a) + a \cdot \tilde{I}] = r_\gamma(i, b^i) + b^i \cdot \tilde{I}, \ i = 1, \ldots, m,$$

then $b^i = a^{(r,i)}, i = 1, \ldots, m$.

Since the matrix $\tilde{M}^k$ is positive there exists $\Delta(r, \gamma) > 0$ such that the following property holds:

(P2)   if $M \in \mathcal{M}$ satisfies $|M - \tilde{M}|_\infty \leq \Delta(r, \gamma)$, then M is k-positive.

In view of (P1), (P2) and Proposition 2.11, there exists an open neighborhood $\mathcal{U}(r, \gamma)$ of r_γ in $\mathcal{A}$ such that the following property holds:

(P3)   If

$$h \in \mathcal{U}(r, \gamma), \ I \in R^m, \ b^i \in A(i), \ i = 1, \ldots, m,$$

$$I_i + \lambda(h) = \min_{a \in A(i)} [h(i, a) + a \cdot I] = h(i, b^i) + b^i \cdot I, \ i = 1, \ldots, m,$$

then

$$\|b^i - a^{(r,i)}\| \leq \Delta(r, \gamma)/2, \ i = 1, \ldots, m$$

and the matrix D with $D_i = b^i, i = 1, \ldots, m$, is k-positive.

Define

$$\mathcal{F}_0 = \cup \{\mathcal{U}(r, \gamma) : \ r \in \mathcal{A}, \ \gamma \in (0, 1)\}. \tag{3.8}$$

Clearly, $\mathcal{F}_0$ is an open everywhere dense subset of $\mathcal{A}$.
Assume that

$$h \in \mathcal{F}_0, \ I \in R^m, \ h_i \in A(i), \ i = 1, \ldots, m, \tag{3.9}$$

$$I_i + \lambda(h) = \min_{a \in A(i)} [h(i, a) + a \cdot I] = h(i, b^i) + b^i \cdot I, \ i = 1, \ldots, m. \tag{3.10}$$

By (3.8) and (3.9), there exist $r \in \mathcal{A}, \gamma \in (0, 1)$ such that

$$h \in \mathcal{U}(r, \gamma). \tag{3.11}$$

In view of (3.9)–(3.11), the matrix D with $D_i = b^i, i = 1, \ldots, m$ is k-positive where k is a natural number.

Thus we have shown the existence of open everywhere dense set $\mathcal{F}_0 \subset \mathcal{A}$ such that the following property holds:
if (3.9) and (3.10) hold, then the matrix D with $D_i = b^i, i = 1, \ldots, m$ is k-positive where k is a natural number.

Together with Theorem 2.7 this completes the proof of Theorem 3.3.

3.2   **Good Policies**

Let $r \in C_{reg}(G)$. Then there exists $I = (I_1, \ldots, I_m) \in R^n$ such that for each $i \in \{1, \ldots, m\}$,

$$I_i + \lambda(r) = \min_{a \in A(i)} [r(i, a) + a \cdot I]. \tag{3.12}$$

By (3.12) for each $i \in \{1, \ldots, m\}$ there exists $\widehat{a}^i \in A(i)$ such that

$$I_i + \lambda(r) = r(i, \widehat{a}^i) + \widehat{a}^i \cdot I. \tag{3.13}$$

For all $i \in \{1, \ldots, m\}$ and all $a \in A(i)$ set

$$\theta(i, a) = r(i, a) + a \cdot I - I_i - \lambda(r). \tag{3.14}$$

By (3.12)–(3.14),

$$\theta(i, a) \geq 0, \ i \in \{1, \ldots, m\}, \ a \in A(i), \tag{3.15}$$

$$\theta(i, \widehat{a}^i) = 0, \ i \in \{1, \ldots, m\}. \tag{3.16}$$

Consider the matrix $\widehat{M} \in \mathcal{M}$ such that

$$\widehat{M}_i = \widehat{a}_i, \ i = 1, \ldots, m \tag{3.17}$$

and associate with $\widehat{M}$ the nonrandomized stationary Markovian policy $\widehat{\sigma} = \{\widehat{M}\}^\infty$.

Proposition 3.5 *Let $x^{(0)} \in A$, let $\sigma = \{M^{(k)}\}_{k=0}^\infty$ be any Markov policy and let*

$$x^{(k+1)} = (M^{(k)})^T x^{(k)}, \ k = 0, 1, \ldots. \tag{3.18}$$

Then for each natural number N,

$$C_N^r(x^{(0)}, \sigma) = \sum_{k=0}^{N-1} \sum_{j=1}^{m} x_j^{(k)} r(j, M_j^{(k)})$$

$$= \lambda(r)N + \sum_{k=0}^{N-1} \sum_{j=1}^{m} x_j^{(k)} \theta(j, M_j^{(k)}) + x^{(0)} \cdot I - x^{(N)} \cdot I.$$

Proof Let $N \geq 1$ be an integer. By (1.2), (3.14) and (3.16),

$$C_N^{r_\gamma}(x^{(0)}, \sigma) = \sum_{k=0}^{N-1} \sum_{j=1}^{m} x_j^{(k)} r(j, M_j^{(k)})$$

$$= \sum_{k=0}^{N-1} \sum_{j=1}^{m} x_j^{(k)} (\theta(j, M_j^{(k)}) + I_j + \lambda(r) - M_j^{(k)} \cdot I)$$

$$= \lambda(r)N + \sum_{k=0}^{N-1} \sum_{j=1}^{m} x_j^{(k)} \theta(j, M_j^{(k)}) + \sum_{k=0}^{N-1} x^{(k)} \cdot I - \sum_{k=0}^{N-1} \sum_{j=1}^{m} \sum_{q=1}^{m} x_j^{(k)} I_q M_{j,q}^{(k)}$$

$$= \lambda(r)N + \sum_{k=0}^{N-1} \sum_{j=1}^{m} x_j^{(k)} \theta(j, M_j^{(k)}) + \sum_{k=0}^{N-1} x^{(k)} \cdot I$$

$$- \sum_{k=0}^{N-1} \sum_{q=1}^{m} [(\sum_{j=1}^{m} x_j^{(k)} M_{j,q}^{(k)}) I_q]$$

$$= \lambda N(r) + \sum_{k=0}^{N-1} \sum_{j=1}^{m} x_j^{(k)} \theta(j, M_j^{(k)})$$

$$+ \sum_{k=0}^{N-1} \sum_{j=1}^{m} x_j^{(k)} I_j - \sum_{k=0}^{N-1} \sum_{q=1}^{m} x_q^{(k+1)} I_q$$

$$= \lambda(r)N + \sum_{k=0}^{N-1} \sum_{j=1}^{m} x_j^{(k)} \theta(j, M_j^{(k)}) + x^{(0)} \cdot I - x^{(N)} \cdot I.$$

Proposition 3.5 is proved.

Proposition 3.5 implies the following result.

Proposition 3.6 *Let $x^{(0)} \in A$, let $\sigma = \{M^{(k)}\}_{k=0}^{\infty}$ be any Markov policy and let*

$$x^{(k+1)} = (M^{(k)})^T x^{(k)}, \ k = 0, 1, \ldots.$$

Then either the sequence $\{C_N^r(x^{(0)}, \sigma) - N\lambda(r)\}_{N=1}^{\infty}$ is bounded or

$$\lim_{N \to \infty} (C_N^r(x^{(0)}, \sigma) - N\lambda(r)) = \infty.$$

Moreover, the sequence $\{C_N^r(x^{(0)}, \sigma) - N\lambda(r)\}_{N=1}^{\infty}$ *is bounded or if and only if*

$$\sum_{k=0}^{\infty} \left(\sum_{j=1}^{m} x_j^{(k)} \theta(j, M_j^{(k)}) \right) < \infty.$$

Let $x^{(0)} \in A$ and let $\sigma = \{M^{(k)}\}_{k=0}^{\infty}$ be a Markov policy. A pair $(x^{(0)}, \sigma)$ is called good if the sequence $\{C_N^r(x^{(0)}, \sigma) - N\lambda(r)\}_{N=1}^{\infty}$ is bounded.

Proposition 3.5 implies that $(x^{(0)}, \widehat{\sigma})$ is good for every $x^{(0)} \in \mathcal{M}$.

Let $x^{(0)} \in A$, let $\sigma = \{M^{(k)}\}_{k=0}^{\infty}$ be a Markov policy and let

$$x^{(k+1)} = (M^{(k)})^T x^{(k)}, \quad k = 0, 1, \ldots.$$

Denote by $\Omega(x^{(0)}, \sigma)$ the set of all pairs $(y, M) \in A \times \mathcal{M}$ for which there exists a strictly increasing sequence of natural numbers $\{k_j\}_{j=1}^{\infty}$ such that

$$y = \lim_{j \to \infty} x^{(k_j)}, \quad M = \lim_{j \to \infty} M^{(k_j)}.$$

Proposition 3.6 implies the following result.

Proposition 3.7 *Let* $x^{(0)} \in A$, $\sigma = \{M^{(k)}\}_{k=0}^{\infty}$ *be a Markov policy, a pair* $(x^{(0)}, \sigma)$ *be good and let* $(y, M) \in \Omega(x^{(0)}, \sigma)$. *Then*

$$\sum_{i=1}^{m} y_i \theta(i, M_i) = 0.$$

Proposition 3.8 *Let* $x^{(0)} \in A$, $\sigma = \{M^{(k)}\}_{k=0}^{\infty}$ *be a Markov policy and let* $(y, Q) \in \Omega(x^{(0)}, \sigma)$. *Then there exists a sequence* $\{(y^{(k)}, Q^{(k)})\}_{k=-\infty}^{\infty} \subset \Omega(x^{(0)}, \sigma)$ *such that for each integer* k,

$$y^{(k+1)} = (Q^{(k)})^T y^{(k)}, \quad y^{(0)} = y, \quad Q^{(0)} = Q.$$

Proof There exists a strictly increasing sequence of natural numbers $\{k_j\}_{j=1}^{\infty}$ such that

$$Q = \lim_{j \to \infty} M^{(k_j)}, \quad y = \lim_{j \to \infty} x^{(k_j)}, \tag{3.19}$$

where

$$x^{(k+1)} = (M^{(k)})^T x^{(k)}, \quad k = 0, 1, \ldots. \tag{3.20}$$

For any natural number j set

$$y^{(j,p)} = x^{p+k_j}, \quad Q^{(j,p)} = M^{p+k_j} \text{ for all integers } p \geq -k_j. \tag{3.21}$$

Extracting subsequences and using a diagonalization process we obtain that there exists a strictly increasing sequence of natural numbers $\{j_s\}_{s=1}^{\infty}$ such that for every integer p there exist

$$y^{(p)} = \lim_{s \to \infty} y^{(j_s,p)}, \quad Q^{(p)} = \lim_{s \to \infty} Q^{(j_s,p)}.$$

It is easy to see that

$$y^{(0)} = y, \quad Q^{(0)} = Q,$$

$$(y^{(p)}, Q^{(p))}\} \in \Omega(x^{(0)}, \sigma)$$

for all integers p and that

$$y^{(p+1)} = (Q^{(p)})^T y^{(p)}$$

for all integers p. Proposition 3.8 is proved.

We will prove the following result which shows that the asymptotic turnpike property holds with the pair $(z(\widehat{M}), \widehat{M})$ being the turnpike.

Theorem 3.9 *Assume that the matrix $\widehat{M}$ is $\tilde{l}$-positive where $\tilde{l}$ is a natural number and if $i \in \{1, \ldots, m\}$ and $b \in A(i)$ satisfies*

$$I_i + \lambda(r) = r(i, b) + b \cdot I,$$

then $b = \widehat{a}^i$. Let $x^{(0)} \in A$, $\sigma = \{M^{(k)}\}_{k=0}^{\infty}$ be a Markov policy, $(x^{(0)}, \sigma)$ be a good pair and let

$$x^{(k+1)} = (M^{(k)})^T x^{(k)}, \quad k = 0, 1, \ldots.$$

Then

$$\lim_{k \to \infty} M^{(k)} = \widehat{M}, \quad \lim_{k \to \infty} x^{(k)} = z(\widehat{M}).$$

Note that if all the sets $A(i)$, $i = 1, \ldots, m$ are finite, then the set $\mathcal{M}$ is finite too and in this case by Theorem 3.9 $M^{(k)} = \widehat{M}$ for all sufficiently large natural numbers k.

Let $x^{(0)} \in A$ and $\sigma = \{M^{(k)}\}_{k=0}^{\infty}$ be a Markov policy. A pair $(x^{(0)}, \sigma)$ is overtaking optimal for any other Markov police $\tilde{\sigma} = \{\tilde{M}^{(k)}\}_{k=0}^{\infty}$,

$$\limsup_{N\to\infty}[C_N^r(x^{(0)},\sigma) - C_N^r(x^{(0)},\tilde{\sigma})] \leq 0.$$

We will prove the following result which shows that for every initial state $x^{(0)}$, $(x^{(0)},\widehat{\sigma})$ is a unique overtaking optimal pair.

Theorem 3.10 *Assume that the matrix $\widehat{M}$ is $\tilde{l}$-positive where $\tilde{l}$ is a natural number and if $i \in \{1,\ldots,m\}$ and $b \in A(i)$ satisfies*

$$I_i + \lambda(r) = r(i,b) + b \cdot I,$$

then $b = \widehat{a}^i$. Then for each $x^{(0)} \in A$ the pair $(x^{(0)},\widehat{\sigma})$ is overtaking optimal, where $\widehat{\sigma} = \{\widehat{M}\}^\infty$. Moreover, if $x^{(0)} \in A$, $\sigma = \{M^{(k)}\}_{k=0}^\infty$ is a Markov policy,

$$x^{(k+1)} = (M^{(k)})^T x^{(k)}, \ k = 0, 1, \ldots$$

and $(x^{(0)},\sigma)$ is an overtaking optimal pair, then

$$x^{(k+1)} = (\widehat{M})^T x^{(k)}, \ k = 0, 1, \ldots,$$

$$\lim_{k\to\infty} x^{(k)} = \widehat{z}(M)$$

and if $k \geq 0$ is an integer and $x^{(k)} >> 0$, then $M^{(k)}) = \widehat{M}$.

3.3   Proof of Theorem 3.9

Since the matrix $\widehat{M}$ is $\tilde{l}$-positive we have

$$(\widehat{M}^T)^{(k)}(x) \to z(\widehat{M}) \text{ as } k \to \infty \tag{3.22}$$

uniformly on A. Let $(y, Q) \in \Omega(x^{(0)},\sigma)$. In order to prove Theorem 3.9 it is sufficient to show that

$$y = z(\widehat{M}), \ Q = \widehat{M}.$$

Proposition 3.8 implies that there exists a sequence

$$\{y^{(k)}, Q^{(k)}\}_{k=-\infty}^\infty \subset \Omega(x^{(0)},\sigma) \tag{3.23}$$

such that

$$y^{(0)} = y, \ Q^{(0)} = Q \tag{3.24}$$

and that for all integers k,

$$y^{(k+1)} = (Q^{(k)})^T y^{(k)}. \tag{3.25}$$

Proposition 3.7 and (3.23) imply that for each integer k,

$$\sum_{i=1}^{m} y_i^{(k)} \theta(i, Q_i^{(k)}) = 0. \tag{3.26}$$

In view of (3.12)–(3.14) and the assumptions of the theorem,

$$\text{if } i \in \{1, \ldots, m\}, \ b \in A(i) \text{ and } \theta(i, b) = 0, \text{ then } b = \widehat{a}_i. \tag{3.27}$$

Let k be an integer. By (3.26) and (3.27),

$$\text{if } i \in \{1, \ldots, m\} \text{ and } y_i^{(k)} > 0, \text{ then } Q_i^{(k)} = \widehat{a}^{(i)}. \tag{3.28}$$

We show that

$$y^{(k+1)} = (\widehat{M})^T y^{(k)}.$$

Set

$$E = \{i \in \{1, \ldots, m\} : \ y_i^{(k)} > 0\}. \tag{3.29}$$

By (3.28) and (3.29),

$$Q_i^{(k)} = \widehat{a}^{(i)}, \ i \in E. \tag{3.30}$$

It follows from (3.16), (3.25), (3.27), (3.29) and (3.30) that for any $j \in \{1, \ldots, m\}$,

$$y_j^{(k+1)} = \sum_{i=1}^{m} y_i^{(k)} Q_{i,j}^{(k)} = \sum_{i \in E} y_i^{(k)} Q_{i,j}^{(k)}$$

$$= \sum_{i \in E} y_i^{(k)} \widehat{a}_j^{(i)} = \sum_{i=1}^{m} y_i^{(k)} \widehat{a}_j^{(i)} = ((\widehat{M})^T y^{(k)})_j.$$

Thus

$$y^{(k+1)} = (\widehat{M})^T y^{(k)}$$

for all integers k. Together with (3.21) this implies that

$$y^{(k)} = z(\widehat{M})$$

for all integers k. Combined with (3.27) and (3.28) this implies that for all integers k,

$$Q_i^{(k)} = \widehat{a}^{(i)}, \quad i = 1, \ldots, m, \quad Q^{(k)} = \widehat{M}$$

and in particular

$$y = y^{(0)} = z(\widehat{M}), \quad Q = Q^{(0)} = \widehat{M}.$$

This completes the proof of Theorem 3.9.

3.4   Proof of Theorem 3.10

In view of the assumptions of the theorem the following property holds:

$$\text{if } i \in \{1, \ldots, m\}, \ b \in A(i), \ \theta(i, b) = 0, \quad \text{then } b = \widehat{a}^i = \widehat{M}_i. \tag{3.31}$$

Let $x^{(0)} \in A$. Consider a stationary Markov police $\widehat{\sigma} = \{\widehat{M}\}^{\infty}$ and set

$$\widehat{x}^{(0)} = x^{(0)}, \quad \widehat{x}^{(k+1)} = (\widehat{M})^T \widehat{x}^{(k)}, \quad k = 0, 1, \ldots. \tag{3.32}$$

Proposition 3.5 and (3.14), (3.17), (3.31) and (3.32) imply that

$$C_N^r(x^{(0)}, \widehat{\sigma}) = \lambda(r)N + \sum_{k=0}^{N-1} \sum_{i=1}^{m} \widehat{x}_i^{(k)} \theta(i, \widehat{M}_i) + x^{(0)} \cdot I - \widehat{x}^{(N)} \cdot I$$

$$= \lambda(r)N + x^{(0)} \cdot I - \widehat{x}^{(N)} \cdot I. \tag{3.33}$$

In view of (3.33), the pair $(x^{(0)}, \widehat{\sigma})$ is good. Theorem 3.9 implies that

$$\lim_{N \to \infty} \widehat{x}^{(N)} = z(\widehat{M}). \tag{3.34}$$

Assume that $\sigma = \{M^{(k)}\}_{k=0}^{\infty}$ is a Markov police and that for each integer $k \geq 0$,

$$x^{(k+1)} = (M^{(k)})^T x^{(k)}. \tag{3.35}$$

If $(x^{(0)}, \sigma)$ is not good, then

$$\lim_{N \to \infty} [C_N^r(x^{(0)}, \widehat{\sigma}) - C_N^r(x^{(0)}, \sigma)] = -\infty.$$

Assume that $(x^{(0)}, \sigma)$ is good. Theorem 3.9 and (3.35) imply that

$$\lim_{N \to \infty} x^{(N)} = z(\widehat{M}). \tag{3.36}$$

Proposition 3.5 and (3.35) imply that for any natural number N,

$$C_N^r(x^{(0)}, \sigma) = \lambda(r)N + \sum_{k=0}^{N-1} \sum_{i=1}^{m} x_i^{(k)} \theta(i, M_i^{(k)}) + x^{(0)} \cdot I - x^{(N)} \cdot I. \tag{3.37}$$

Since the pair $(x^{(0)}, \sigma)$ is good we have in view of (3.37),

$$\sum_{k=0}^{\infty} \sum_{i=1}^{m} x_i^{(k)} \theta(i, M_i^{(k)}) < \infty$$

and by (3.36) and (3.37),

$$\lim_{N \to \infty} [C_N^r(x^{(0)}, \sigma) - \lambda(r)N]$$

$$= \sum_{k=0}^{\infty} \sum_{i=1}^{m} x_i^{(k)} \theta(i, M_i^{(k)}) + x^{(0)} \cdot I - z(\widehat{M}) \cdot I. \tag{3.38}$$

Equations (3.33) and (3.34) imply that

$$\lim_{N \to \infty} [C_N^r(x^{(0)}, \widehat{\sigma}) - \lambda(r)N] = x^{(0)} \cdot I - z(\widehat{M}) \cdot I. \tag{3.39}$$

It follows from (3.38) and (3.39) that

$$\lim_{N \to \infty} [C_N^r(x^{(0)}, \widehat{\sigma}) - C_N^r(x^{(0)}, \sigma)]$$

$$= -\sum_{k=0}^{\infty} \sum_{i=1}^{m} x_i^{(k)} \theta(i, M_i^{(k)}) \leq 0. \tag{3.40}$$

Thus $(x^{(0)}, \widehat{\sigma})$ is overtaking optimal.

Assume that $(x^{(0)}, \sigma)$ is also an overtaking optimal pair. Then in view of (3.40),

$$\sum_{k=0}^{\infty} \sum_{i=1}^{m} x_i^{(k)} \theta(i, M_i^{(k)}) = 0. \tag{3.41}$$

Let $k \geq 0$ be an integer. Set

$$E_k = \{i \in \{1, \ldots, m\} : x_i^{(k)} > 0\}. \tag{3.42}$$

By (3.31), (3.41) and (3.42), for each $i \in E_k$,

$$\theta(i, M_i^{(k)}) = 0, \quad M_i^{(k)} = \widehat{a}^i = \widehat{M}_i. \tag{3.43}$$

In view of (3.35), (3.42) and (3.43), for each $j \in \{1, \ldots, m\}$,

$$x_j^{(k+1)} = \sum_{i=1}^{m} x_i^{(k)} M_{i,j}^{(k)} = \sum_{i \in E_k} x_i^{(k)} M_{i,j}^{(k)} = \sum_{i \in E_k} x_i^{(k)} \widehat{M}_{i,j} = \sum_{i=1}^{m} x_i^{(k)} \widehat{M}_{i,j}$$

and

$$x^{(k+1)} = (\widehat{M})^T x^{(k)}$$

for all integers $k \geq 0$. Since $\widehat{M}$ is $\tilde{l}$-positive we conclude that

$$\lim_{k \to \infty} x^{(k)} = z(\widehat{M}).$$

Assume that an integer $k \geq 0$ and that

$$x^{(k)} >> 0.$$

Together with (3.31) and (3.41) this implies that for all $i \in \{1, \ldots, m\}$,

$$\theta(i, M_i^{(k)}) = 0, \quad M_i^{(k)} = \widehat{a}^i = \widehat{M}_i$$

and $M^{(k)} = \widehat{M}$. Theorem 3.10 is proved.

Chapter 4
A Weak Turnpike Property

In this chapter we consider the class of (MDPs) studied in Chap. 3 and assuming the uniqueness of minimizing Markov actions establish the weak turnpike property for approximate optimal programs on finite horizon.

4.1 A Turnpike Result

In this chapter we use the notation and definitions introduced in Sects. 2.1 2.3 and 3.1, 3.2.

Assume that $r \in C_{reg}(G)$. Then there exists $I = (I_1, \ldots, I_m) \in R^n$ such that for each $i \in \{1, \ldots, m\}$,

$$I_i + \lambda(r) = \min_{a \in A(i)} [r(i, a) + a \cdot I]. \tag{4.1}$$

By (4.1) for each $i \in \{1, \ldots, m\}$ there exists

$$\widehat{a}^i \in A(i) \tag{4.2}$$

such that

$$I_i + \lambda(r) = r(i, \widehat{a}^i) + \widehat{a}^i \cdot I. \tag{4.3}$$

For all $i \in \{1, \ldots, m\}$ and all $a \in A(i)$ set

$$\theta(i, a) = r(i, a) + a \cdot I - I_i - \lambda(r). \tag{4.4}$$

A. J. Zaslavski, *Turnpike Phenomenon for Markov Decision Processes*,
SpringerBriefs in Mathematics, https://doi.org/10.1007/978-3-032-00854-1_4

By (4.1)–(4.3),

$$\theta(i,a) \ge 0, \ i \in \{1,\dots,m\}, \ a \in A(i), \tag{4.5}$$

$$\theta(i,\widehat{a}^i) = 0, \ i \in \{1,\dots,m\}. \tag{4.6}$$

Consider the matrix $\widehat{M} \in \mathcal{M}$ such that

$$\widehat{M}_i = \widehat{a}_i, \ i = 1,\dots,m \tag{4.7}$$

and associate with $\widehat{M}$ the nonrandomized stationary Markovian policy $\widehat{\sigma} = \{\widehat{M}\}^\infty$. Proposition 2.1 and Eqs. (4.6), (4.7) imply that for each $x^{(0)} \in A$ and each natural number N,

$$|C_N^r(x^{(0)},\widehat{\sigma}) - \lambda(r)N| \le 2|I|_\infty. \tag{4.8}$$

We suppose that if $i \in \{1,\dots,m\}$ and $b \in A(i)$ satisfies

$$I_i + \lambda(r) = r(i,b) + b \cdot I, \quad \text{then } b = \widehat{a}^i. \tag{4.9}$$

Equations (4.4) and (4.9) imply that

$$\text{if } i \in \{1,\dots,m\}, \ b \in A(i), \ \theta(i,b) = 0, \ \text{then } b = \widehat{a}^i. \tag{4.10}$$

We prove the following result which shows that approximate optimal solutions on large intervals possess a weak turnpike property.

Theorem 4.1 *Assume that $\tilde{l}$ is a natural number, $\widehat{M}$ is $\tilde{l}$-positive and that $\epsilon, c_0 > 0$. Then there exists a natural number L_1 such that for each $x^{(0)} \in A$ and each Markov police $\sigma = \{M^{(k)}\}_{k=0}^\infty \subset \mathcal{M}$ with*

$$x^{(k+1)} = (M^{(k)})^T x^{(k)}, \ k = 0, 1, \dots$$

and each natural number $N \ge L_1$ such that

$$C_N^r(x^{(0)},\sigma) \le \lambda(r)N + c_0$$

the inequality

$$Card(\{k \in \{0,\dots,N-1\}: \ \|x^{(k)} - z(\widehat{M})\| + |M^{(k)} - \widehat{M}|_\infty > \epsilon\}) \le L_1$$

holds.

Note that if all the sets $A(i), i = 1,\dots,m$ are finite, then the set $\mathcal{M}$ is finite too and in this case by Theorem 4.1 for small enough ϵ, $M^{(k)} = \widehat{M}$ for all sufficiently large natural numbers k.

4.2 Auxiliary Results for Theorem 4.1

Since the matrix $\widehat{M}$ is $\tilde{l}$ positive we have

$$(\widehat{M}^T)^{(k)}(x) \to z(\widehat{M}) \text{ as } k \to \infty \tag{4.11}$$

uniformly on A.

Lemma 4.2 *Let* $\{M^{(k)}\}_{k=-\infty}^{-1} \subset \mathcal{M}$, $\{x^{(k)}\}_{k=-\infty}^{0} \subset \mathcal{A}$,

$$x^{(k+1)} = (M^{(k)})^T x^{(k)} \text{ for all integers } k < 0, \tag{4.12}$$

$$\sum_{i=1}^{m} x_i^{(k)} \theta(i, M_i^{(k)}) = 0 \text{ for all integers } k < 0. \tag{4.13}$$

Then $x^{(k)} = z(\widehat{M})$ *for all integers* $k \le 0$ *and* $M^{(k)} = \widehat{M}$ *for all integers* $k < 0$.

Proof Let an integer $k < 0$. Set

$$E = \{i \in \{1, \ldots, m\} : \ x_i^{(k)} > 0\}. \tag{4.14}$$

By (4.7), (4.9), (4.13) and (4.14), for all $i \in E$,

$$\theta(i, M_i^{(k)}) = 0, \ M_i^{(k)} = \widehat{a}^i = \widehat{M}_i. \tag{4.15}$$

By (4.12), (4.14) and (4.15), for all $j \in \{1, \ldots, m\}$,

$$x_j^{(k+1)} = \sum_{i=1}^{m} x_i^{(k)} M_{i,j}^{(k)} = \sum_{i \in E} x_i^{(k)} M_{i,j}^{(k)}$$

$$= \sum_{i \in E} x_i^{(k)} \widehat{M}_{i,j} = \sum_{i=1}^{m} x_i^{(k)} \widehat{M}_{i,j}$$

and

$$x^{(k+1)} = (\widehat{M}^T) x^{(k)} \text{ for all integers } k < 0.$$

Together with (4.11) this implies that

$$x^{(k)} = z(\widehat{M}) \text{ for all integers } k \le 0.$$

Together with (4.7), (4.9) and (4.13), this implies that for all integers $k < 0$ and all $i \in \{1, \ldots, m\}$,

$$\theta(i, M_i^{(k)}) = 0, \quad M_i^{(k)} = \widehat{a}^i = \widehat{M}_i, \quad M^{(k)} = \widehat{M}.$$

Lemma 4.2 is proved.

Lemma 4.3 *Let $\epsilon > 0$. Then there exist $\delta > 0$ and a natural number N such that for each $x^{(0)} \in A$, each Markov policy $\{M^{(k)}\}_{k=0}^{\infty} \subset \mathcal{M}$ with*

$$x^{(k+1)} = (M^{(k)})^T x^{(k)}, \quad k = 0, 1, \ldots$$

and a natural number $N > N_1$ such that

$$\sum_{k=0}^{N-1} \sum_{i=1}^{m} x_i^{(k)} \theta(i, M_i^{(k)}) \leq \delta$$

the following inequalities hold:

$$\|x^{(k)} - z(\widehat{M})\| \leq \epsilon, \quad k = N_1, \ldots, N - 1,$$

$$|M^{(k)} - \widehat{M}\|_{\infty} \leq \epsilon, \quad k = N_1, \ldots, N - 1.$$

Proof Assume that the lemma does not hold. Then for each natural number p there exist $x^{(p,0)} \in A$, $\{M^{(p,k)}\}_{k=0}^{\infty} \subset \mathcal{M}$ with

$$x^{(p,k+1)} = (M^{(p,k)})^T x^{(p,k)}, \quad k = 0, 1, \ldots, \tag{4.16}$$

a natural number $N_p > p$ such that

$$\sum_{k=0}^{N_p-1} \sum_{i=1}^{m} x_i^{(p,k)} \theta(i, M_i^{(p,k)}) \leq p^{-1} \tag{4.17}$$

and an integer

$$s_p \in [p, N_p - 1] \tag{4.18}$$

such that

$$\|x^{(p,s_p)} - z(\widehat{M})\| + |M^{(p,k)} - \widehat{M}|_{\infty} > \epsilon. \tag{4.19}$$

Let p be a natural number. Set

$$y^{(p,k)} = x^{(p,k+s_p+1)} \text{ for all integers } k \geq -s_p - 1,$$

$$Q^{(p,k)} = M^{(p,k+s_p+1)} \text{ for all integers } k \geq -s_p - 1. \tag{4.20}$$

By (4.16) and (4.20), for all integers $k \geq -s_p - 1$,

$$y^{(p,k+1)} = (M^{(p,k+s_p+1)})^T x^{(p,k+s_p+1)} = (Q^{(p,k)})^T y^{(p,k)}. \tag{4.21}$$

In view of (4.17) and (4.20), for all integers $k = -1 - s_p, \ldots, N_p - 2 - s_p$,

$$\sum_{i=1}^{m} y_i^{(p,k)} \theta(i, Q_i^{(p,k)}) = \sum_{i=1}^{m} x_i^{(p,k+s_p+1)} \theta(i, M_i^{(p,k+s_p+1)}) \leq p^{-1}. \tag{4.22}$$

Equations (4.19) and (4.20) imply that

$$\|y^{(p,-1)} - z(\widehat{M})\| + |Q^{(p,-1)} - \widehat{M}|_\infty > \epsilon. \tag{4.23}$$

Extracting subsequences, using a diagonalization process and re-indexing we may assume without loss of generality that for every integer $k \leq 0$ there exist

$$y^{(k)} = \lim_{p\to\infty} y^{(p,k)}, \quad Q^{(k)} = \lim_{p\to\infty} Q^{(p,k)}. \tag{4.24}$$

By (4.21) and (4.24), for each integer $k < 0$.

$$y^{(k+1)} = (Q^{(k)})^T y^{(k)}. \tag{4.25}$$

It follows from (4.22) and (4.24) that for each integer $k < 0$,

$$\sum_{i=1}^{m} y_i^{(k)} \theta(i, Q_i^{(k)}) = 0. \tag{4.26}$$

In view of (4.23) and (4.24),

$$\|y^{(-1)} - z(\widehat{M})\| + |Q^{(-1)} - \widehat{M}|_\infty \geq \epsilon/2. \tag{4.27}$$

Lemma 4.2 and Eqs. (4.25) and (4.26) imply that $y^{(k)} = z(\widehat{M})$ for all integers $k \leq 0$ and $Q^{(k)} = \widehat{M}$ for all integers $k < 0$. This contradicts (4.27). The contradiction we have reached proves Lemma 4.3.

4.3   Proof of Theorem 4.1

By Lemma 4.3, there exist $\delta > 0$ and a natural number N_1 such that the following property holds:

(P1)   for each $x^{(0)} \in A$, each Markov policy $\{M^{(k)}\}_{k=0}^{\infty} \subset \mathcal{M}$ with

$$x^{(k+1)} = (M^{(k)})^T x^{(k)}, \quad k = 0, 1, \ldots$$

and a natural number $N > N_1$ such that

$$\sum_{k=0}^{N-1} \sum_{i=1}^{m} x_i^{(k)} \theta(i, M_i^{(k)}) \leq \delta$$

the following inequalities hold:

$$\|x^{(k)} - z(\widehat{M})\| \leq \epsilon/4, \quad k = N_1, \ldots, N-1,$$

$$|M^{(k)} - \widehat{M}\|_{\infty} \leq \epsilon/4, \quad k = N_1, \ldots, N-1.$$

Choose a natural number

$$L_1 \geq (2N_1 + 7)(\delta^{-1}(c_0 + 2|I|_{\infty}) + 1). \tag{4.28}$$

Assume that

$$x^{(0)} \in A, \quad \{M^{(k)}\}_{k=0}^{\infty} \subset \mathcal{M},$$

$$x^{(k+1)} = (M^{(k)})^T x^{(k)}, \quad k = 0, 1, \ldots, \tag{4.29}$$

a natural number $N \geq L_1$ and that

$$C_N^r(x^{(0)}, \sigma) \leq \lambda(r)N + c_0. \tag{4.30}$$

Proposition 3.5 and Eqs. (4.29), (4.30) imply that

$$c_0 + \lambda(r)N \geq C_N^r(x^{(0)}, \sigma)$$

$$= \sum_{k=0}^{N-1} \sum_{i=1}^{m} x_i^{(k)} \theta(i, M_i^{(k)}) + x^{(0)} \cdot I - x^{(N)} \cdot I + \lambda(r)N$$

and

$$\sum_{k=0}^{N-1}\sum_{i=1}^{m} x_i^{(k)}\theta(i, M_i^{(k)}) \le c_0 + 2|I|_\infty. \tag{4.31}$$

Set $t_0 = 0$. By induction we will define a finite strictly increasing sequence of integers $t_0, \ldots, t_s$. Assume that $j \ge 0$ is an integer and that we have already defined integers $t_0, \ldots, t_j$. If $t_j = N$, then the construction is completed. Assume that $t_j < N$. If

$$\sum_{k=t_j}^{N-1}\sum_{i=1}^{m} x_i^{(k)}\theta(i, M_i^{(k)}) \le \delta,$$

then we set

$$t_{j+1} = N \tag{4.32}$$

and the construction is completed.

Assume that

$$\sum_{k=t_j}^{N-1}\sum_{i=1}^{m} x_i^{(k)}\theta(i, M_i^{(k)}) > \delta. \tag{4.33}$$

Then there exists a unique integer $t_{j+1} > t_j$ such that $t_{j+1} \le N$,

$$\sum_{k=t_j}^{t_{j+1}-1}\sum_{i=1}^{m} x_i^{(k)}\theta(i, M_i^{(k)}) > \delta \tag{4.34}$$

and for any integer p satisfying $t_j < p < t_{j+1}$,

$$\sum_{k=t_j}^{p-1}\sum_{i=1}^{m} x_i^{(k)}\theta(i, M_i^{(k)}) \le \delta. \tag{4.35}$$

Thus by induction we have constructed a strictly increasing sequence of integers $t_0, t_1, \ldots$ which is finite and such that (4.34), (4.35) hold. Let t_s be its last element. Clearly, $t_s = N$ and $s \ge 1$. By (4.31) and (4.34),

$$c_0 + \lambda(r)N$$

$$\ge \sum\{\sum_{k=t_j}^{t_{j+1}-1}\sum_{i=1}^{m} x_i^{(k)}\theta(i, M_i^{(k)}) : \text{ an integer } j$$

$$\text{satisfies } 0 \le j \le s - 2\} \ge \delta(s - 1)$$

and

$$s \le \delta^{-1}(c_0 + 2|I|_\infty) + 1. \tag{4.36}$$

Define

$$E = \{j \in \{0, \ldots, s - 1\} : \ t_{j+1} - t_j \ge N_1 + 4\}. \tag{4.37}$$

Let

$$j \in E.$$

By (4.37),

$$t_{j+1} - t_j > N_1 + 4. \tag{4.38}$$

and by the definition of t_{j+1}, (4.32) and (4.35),

$$\sum_{k=t_j}^{t_{j+1}-2} \sum_{i=1}^{m} x_i^{(k)} \theta(i, M_i^{(k)}) \le \delta. \tag{4.39}$$

By (4.29), (4.38) and (4.39) and property (P1) applied to the initial distribution x^{t_j} and to the Markov policy $\{M^{(t_j+k)}\}_{k=0}^\infty$), for all integers $k = t_j + N_1, \ldots, t_{j+1} - 2$,

$$\|x^{(k)} - z(\widehat{M})\| \le \epsilon/4, \ \ |M^{(k)} - \widehat{M}|_\infty \le \epsilon/4.$$

This implies that

$$F := \{k \in \{0, \ldots, N - 1\} : \ \|x^{(k)} - z(\widehat{M})\| + |M^{(k)} - \widehat{M}|_\infty > \epsilon\}$$

$$\subset \cup\{\{t_j, \ldots, t_{j+1}\} : \ j \in \{0, \ldots, s - 1\} \setminus E\}$$

$$\cup \{\{t_j, \ldots, t_j + N_1 - 1\} \cup \{\{t_{j+1} - 1, \ t_{j+1}\} : \ j \in E\}. \tag{4.40}$$

By (4.28), (4.36), (4.37) and (4.40)

$$\mathrm{Card}(F) \le s(N_1 + 5) + s(N_1 + 2)$$

$$\le (2N_1 + 7)(\delta^{-1}(c_0 + 2|I|_\infty) + 1) \le L_1.$$

Theorem 4.1 is proved.

Chapter 5
Convex Markov Decision Processes

In this chapter we consider the subclass of MDPs studied in Chap. 2 with convex action sets and convex cost functions. We show that if cost functions are strictly convex, then minimizing Markov actions are unique and that most in the sense of Baire category cost functions are in fact strictly convex.

5.1 Strict Convexity and Generic Results

In this chapter we use the notation and definitions introduced in Sects. 2.1–2.3 and 3.1, 3.2.

Assume that $r \in C_{reg}(G)$. Then there exists $I = (I_1, \ldots, I_m) \in R^m$ such that for each $i \in \{1, \ldots, m\}$,

$$I_i + \lambda(r) = \min_{a \in A(i)} [r(i, a) + a \cdot I]. \tag{5.1}$$

By (5.1) for each $i \in \{1, \ldots, m\}$ there exists

$$\widehat{a}^i \in A(i)$$

such that

$$I_i + \lambda(r) = r(i, \widehat{a}^i) + \widehat{a}^i \cdot I. \tag{5.2}$$

For all $i \in \{1, \ldots, m\}$ and all $a \in A(i)$ set

$$\theta(i, a) = r(i, a) + a \cdot I - I_i - \lambda(r). \tag{5.3}$$

A. J. Zaslavski, *Turnpike Phenomenon for Markov Decision Processes*,
SpringerBriefs in Mathematics, https://doi.org/10.1007/978-3-032-00854-1_5

By (5.1)–(5.3),

$$\theta(i, a) \geq 0, \ i \in \{1, \ldots, m\}, \ a \in A(i), \tag{5.4}$$

$$\theta(i, \widehat{a}^i) = 0, \ i \in \{1, \ldots, m\}. \tag{5.5}$$

Consider the matrix $\widehat{M} \in \mathcal{M}$ such that

$$\widehat{M}_i = \widehat{a}_i, \ i = 1, \ldots, m \tag{5.6}$$

and associate with $\widehat{M}$ the nonrandomized stationary Markovian policy $\widehat{\sigma} = \{\widehat{M}\}^\infty$.
In this chapter we prove the following results.

Theorem 5.1 *Assume that the matrix $\widehat{M}$ is good and that*

$$if \ i \in \{1, \ldots, m\}, \ b \in A(i),$$

$$I_i + \lambda(r) = r(i, b) + b \cdot I, \ then \ b = \widehat{a}^i. \tag{5.7}$$

Then if $\tilde{I} = (\tilde{I}_1, \ldots, \tilde{I}_m) \in R^m$, $b^i \in A(i)$, $i = 1, \ldots, m$ satisfy

$$\tilde{I}_i + \lambda(r) = \min_{a \in A(i)} [r(i, a) + a \cdot \tilde{I}] = r(i, b^i) + b^i \cdot \tilde{I}, \ i = 1, \ldots, m,$$

then $b^i = \widehat{a}^i$, $i = 1, \ldots, m$.

Assume that $E \subset R^m$ is a convex set. A function $f : E \to R^1$ is called strictly
convex [84] if for each $\alpha \in (0, 1)$ and each $x, y \in E$ satisfying $x \neq y$,

$$f(\alpha x + (1 - \alpha)y) < \alpha f(x) + (1 - \alpha)f(y).$$

Theorem 5.1 implies the following result which shows that Eq. (2.1) has a unique
solution if the cost functions are strictly convex.

Theorem 5.2 *Assume that for each $i \in \{1, \ldots, m\}$, the set $A(i)$ is convex and the
function $r(i, \cdot) : A(i) \to R^1$ is strictly convex and the matrix $\widehat{M}$ is good. Then if
$\tilde{I} = (\tilde{I}_1, \ldots, \tilde{I}_m) \in R^m$, $b^i \in A(i)$, $i = 1, \ldots, m$ satisfy*

$$\tilde{I}_i + \lambda(r) = \min_{a \in A(i)} [r(i, a) + a \cdot \tilde{I}] = r(i, b^i) + b^i \cdot \tilde{I}, \ i = 1, \ldots, m,$$

then $b^i = \widehat{a}^i$, $i = 1, \ldots, m$.

The next result shows that most cost functions are strictly convex.

Theorem 5.3 *Assume that for each $i \in \{1, \ldots, m\}$, the set $A(i)$ is convex. Then
there exists a set $\mathcal{F} \subset C_{conv}(G)$ which is a countable intersection of open*

everywhere dense subsets of $C_{conv}(G)$ such that for each $r \in \mathcal{F}$ the functions $r(i, \cdot) : A(i) \to R^1$, $i = 1, \ldots, m$ are strictly convex.

5.2 Proof of Theorem 5.1

Assume that $\tilde{I} = (\tilde{I}_1, \ldots, \tilde{I}_m) \in R^m$, $b^{(i)} \in A(i)$, $i \in \{1, \ldots, m\}$, and that for each $i \in \{1, \ldots, m\}$,

$$\tilde{I}_i + \lambda(r) = \min_{a \in A(i)} [r(i, a) + a \cdot \tilde{I}]$$

$$= r(i, b^{(i)}) + b^{(i)} \cdot \tilde{I}. \tag{5.8}$$

Consider the $m \times m$ matrix M such that

$$M_i = b^{(i)}, \quad i = 1, \ldots, m$$

and associate with M the nonrandomized stationary Markovian policy $\sigma = \{M\}^\infty$. Let $x^{(0)} = m^{-1}(1, 1, \ldots, 1) \in R^m$, $\{\xi^{(k)}\}_{k=0}^\infty$ be a Markov chain associated with the pair $(x^{(0)}, \sigma)$ and

$$x^{(k+1)} = (M^{(k)})^T x^{(k)}, \quad k = 0, 1, \ldots.$$

Proposition 2.1 and Eq. (5.8) imply that for each integer $N \geq 1$,

$$|C_N^r(x^{(0)}, \sigma) - \lambda(r)N| \leq 2|\tilde{I}|_\infty. \tag{5.9}$$

In view of (5.3) and (5.7),

$$\text{if } i \in \{1, \ldots, m\}, \; b \in A(i), \; \theta(i, b) = 0, \; \text{then } b = \widehat{a}^{(i)}. \tag{5.10}$$

Proposition 3.5 and (5.8) imply for each natural number N,

$$C_N^r(x^{(0)}, \sigma) - \lambda(r)N$$

$$= \sum_{k=0}^{N-1} \sum_{j=1}^{m} x_j^{(k)} \theta(j, b^{(j)}) + x^{(0)} \cdot I - x^{(N)} \cdot I.$$

Together with (5.9) this implies that

$$\sum_{k=0}^{\infty} \sum_{j=1}^{m} x_j^{(k)} \theta(j, b^{(j)}) < \infty. \tag{5.11}$$

Denote by S the set of all $j \in \{1, \ldots, m\}$ such that

$$\limsup_{k \to \infty} x_j^{(k)} > 0. \tag{5.12}$$

Clearly, $S \neq \emptyset$. Relations (5.10)–(5.12) imply that for all $j \in S$,

$$\theta(j, b^{(j)}) = 0, \ b^{(j)} = \widehat{a}^{(j)}. \tag{5.13}$$

In view of (5.9) and (5.12), the following property holds:

(P2)   if $j \in S, i \in \{1, \ldots, m\}$ and $\widehat{a}_i^{(j)} > 0$, then $i \in S$.

Since the matrix $\widehat{M}$ is good and $S \neq \emptyset$ property (P2) implies that

$$S = \{1, \ldots, m\}.$$

Together with (5.13) this implies that $b^i = \widehat{a}^i, i = 1, \ldots, m$. Theorem 5.1 is proved.

5.3 Proof of Theorem 5.3

Clearly, the function $\|x\|^2, x \in R^n$ is strictly convex. Assume that $r \in C_{conv}(G)$ and $\gamma \in (0, 1)$. There exist $I^{(r)} \in R^m, a^{(r,i)} \in A(i), i = 1, \ldots, m$ be such that for each $i \in \{1, \ldots, m\}$,

$$I_i^{(r)} + \lambda(r) = \min_{a \in A(i)} \{r(i, a) + a \cdot I^{(r)}\} = r(i, a^{(r,i)}) + a^{(r,i)} \cdot I^{(r)}.$$

Define

$$r_\gamma(i, a) = r(i, a) + \gamma \|a - a^{(r,i)}\|^2, \ (i, a) \in G.$$

Clearly, $r_\gamma \in C_{conv}(G)$ and for each $i \in \{1, \ldots, m\}$ the function $r_\gamma(i, \cdot) : A(i) \to R^1$ is strictly convex. Let q be a natural number and $i \in \{1, \ldots, m\}$. Evidently the function

$$\alpha r_\gamma(i, x) + (1 - \alpha) r_\gamma(i, y) - r_\gamma(i, \alpha x + (1 - \alpha)y)$$

is continuous and positive-valued on the set

$$K_{i,q} := \{(x, y, \alpha) \in A(i) \times A(i) \times [0, 1] :$$

$$\|x - y\| \geq (4q)^{-1}, \ \alpha \in [(4q)^{-1}, 1 - (4q)^{-1}]\}.$$

Therefore there exists a number $\Delta(r, \gamma, q) > 0$ such that for each $i \in \{1, \ldots, m\}$ and each $(x, y, \alpha) \in K_{i,q}$,

$$\alpha r_\gamma(i, x) + (1 - \alpha)\alpha r_\gamma(i, y) - r_\gamma(i, \alpha x + (1 - \alpha)y) \geq \Delta(r, \gamma, q).$$

It is easy to see that there exists an open neighborhood $\mathcal{U}(r, \gamma, q)$ of r_γ in $C_{conc}(G)$ such that the following properties hold:

(P3)   for each $i \in \{1, \ldots, m\}$, each $(x, y, \alpha) \in K_{i,q}$ and each $h \in \mathcal{U}(r, \gamma, q)$,

$$\alpha h(i, x) + (1 - \alpha)h(i, y) - h(i, \alpha x + (1 - \alpha)y) \geq \Delta(r, \gamma, q)/2. \qquad (5.14)$$

Define

$$\mathcal{F} = \cap_{q=1}^{\infty} \cup \{\mathcal{U}(r, \gamma, q) : \ r \in C_{conv}(G), \ \gamma \in (0, 1)\}. \qquad (5.15)$$

Clearly, $\mathcal{F}$ is a countable intersection of open everywhere dense subsets of $C_{conv}(G)$.

Assume that

$$h \in \mathcal{F}, \ i \in \{1, \ldots, m\}.$$

We show that the function $h(i, \cdot) : \ A(i) \to R^1$ is strictly convex. Let

$$\alpha \in (0, 1), \ x, y \in A(i), \ x \neq y.$$

Clearly, there exists a natural number q such that

$$(x, y, \alpha) \in K_{i,q}. \qquad (5.16)$$

By (5.15), there exist $r \in C_{conv}(G)$, $\gamma \in (0, 1)$ such that

$$h \in \mathcal{U}(r, \gamma, q).$$

Together with (5.16) and (P3) this implies (5.14) and that $h(i, \cdot) : \ A(i) \to R^1$ is strictly convex. This completes the proof of Theorem 5.3.

Chapter 6
Turnpike Properties for MDPs with Perturbations

In this chapter we assume the uniqueness of minimizing Markov actions and show that the asymptotic turnpike property and the weak turnpike property are stable under the perturbations of cost functions.

6.1 The Asymptotic Turnpike Property

In this chapter we use the notation and definitions introduced in Sects. 2.1–2.3 and 3.1, 3.2.

Assume that $r \in C_{reg}(G)$. Then there exists $I = (I_1, \ldots, I_m) \in R^m$ such that for each $i \in \{1, \ldots, m\}$,

$$I_i + \lambda(r) = \min_{a \in A(i)} [r(i, a) + a \cdot I]. \tag{6.1}$$

Define

$$I(r) = \{I = (I_1, \ldots, I_m) \in R^m :$$

$$(6.1) \text{ holds and } \max\{I_j : \ j \in \{1, \ldots, m\}\} = 0\}. \tag{6.2}$$

In Sect. 2.1 we showed that the function $r \rightarrow \lambda(r)$, $r \in C_{reg}(G)$ is continuous. Moreover Corollary 2.2 implies that the following property holds:

(P4) the function $\lambda(\cdot)$ is Lipschitz on $C_{reg}(G)$ with the Lipschitzian constant 1.

Proposition 2.4 implies that the following property holds:

(P5) Let Ω be a nonempty, bounded subset of $C_{reg}(G)$. Then there exists a constant $K > 0$ such that for each $r \in \Omega$ and each $I \in I(r)$ we have $|I|_\infty \leq K$.

© The Author(s), under exclusive license to Springer Nature Switzerland AG 2025

A. J. Zaslavski, *Turnpike Phenomenon for Markov Decision Processes*,

SpringerBriefs in Mathematics, https://doi.org/10.1007/978-3-032-00854-1_6

Assume $r \in C_{reg}(G)$ and $I^{(r)} \in I(r)$. (Here we fix $I^{(r)}$ but it can be any element of $I(r)$.) Thus for each $i \in \{1, \ldots, m\}$,

$$I_i^{(r)} + \lambda(r) = \min_{a \in A(i)} [r(i, a) + a \cdot I^{(r)}]. \tag{6.3}$$

By (6.3), for each $i \in \{1, \ldots, m\}$, there exists $a^{(r,i)} \in A(i)$ such that for each $i \in \{1, \ldots, m\}$,

$$I_i^{(r)} + \lambda(r) = r(i, a^{(r,i)}) + a^{(r,i)} \cdot I^{(r)}. \tag{6.4}$$

For all $i \in \{1, \ldots, m\}$ and all $a \in A(i)$ set

$$\theta^{(r)}(i, a) = r(i, a) + a \cdot I^{(r)} - I_i^{(r)} - \lambda(r). \tag{6.5}$$

By (6.3)–(6.5),

$$\theta^{(r)}(i, a) \geq 0, \ i \in \{1, \ldots, m\}, \ a \in A(i), \tag{6.6}$$

$$\theta^{(r)}(i, a^{(r,i)}) = 0, \ i \in \{1, \ldots, m\}. \tag{6.7}$$

Consider the matrix $M^{(r)} \in \mathcal{M}$ such that

$$M_i^{(r)} = a^{(r,i)}, \ i = 1, \ldots, m \tag{6.8}$$

and associate with $M^{(r)}$ the nonrandomized stationary Markovian policy $\sigma^{(r)} = \{M^{(r)}\}^{\infty}$.

Assume that $x^{(0)} \in A$ and that $\sigma = \{M^{(k)}\}_{k=0}^{\infty} \subset \mathcal{M}$ is a Markov policy. A pair $(x^{(0)}, \sigma)$ is called (r)-good if the sequence $\{C_N^r(x^{(0)}, \sigma) - N\lambda(r)\}_{N=1}^{\infty}$ is bounded.

We will prove the following result which shows the stability of the asymptotic turnpike property under small perturbations of cost functions.

Theorem 6.1 *Assume that $\bar{r} \in C_{reg}(G)$, $\tilde{l}$ is a natural number, $M^{(\bar{r})}$ is $\tilde{l}$-positive and*

$$\textit{if } \bar{I} \in I(\bar{r}), \ i \in \{1, \ldots, m\}, \ b \in A(i),$$

$$\bar{I}_i + \lambda(\bar{r}) = \bar{r}(i, b) + b \cdot I^{(\bar{r})}, \textit{ then } b = a^{(\bar{r},i)}. \tag{6.9}$$

Let $\epsilon > 0$. Then there exists $\delta > 0$ such that for each $r \in C_{reg}(G)$ which satisfies $\|r - \bar{r}\| \leq \delta$ and each (r)-good pair $(x^{(0)}, \sigma)$ where $\sigma = \{M^{(k)}\}_{k=0}^{\infty} \subset \mathcal{M}, x^{(0)} \in A$ and

$$x^{(k+1)} = (M^{(k)})^T x^{(k)}, \ k = 0, 1, \ldots$$

for all sufficiently large natural number k,

$$\|x^{(k)} - z(M^{(\bar{r})})\| \le \epsilon, \quad |M^{(k)} - M^{(\bar{r})}|_\infty \le \epsilon.$$

Note that if all the sets $A(i)$, $i = 1, \ldots, m$ are finite, then the set $\mathcal{M}$ is finite too and in this case by Theorem 6.1 for small enough ϵ, $M^{(k)} = M^{(\bar{r})}$ for all sufficiently large natural numbers k.

6.2   Auxiliary Results for Theorem 6.1

Lemma 6.2 *Assume*

$$\{r_p\}_{p=1}^\infty \subset C_{reg}(G), \ r \in C_{reg}(G),$$

$$\lim_{p \to \infty} \|r_p - r\| = 0, \tag{6.10}$$

$$I^{(p)} \in I(r_p), \ p = 1, 2, \ldots, \tag{6.11}$$

$$I = \lim_{p \to \infty} I^{(p)}. \tag{6.12}$$

Then $I \in I(r)$.

Proof Clearly,

$$\max\{I_i : i = 1, \ldots, m\} = 0.$$

By (6.10) and the continuity of the functional $\lambda(\cdot)$,

$$\lambda(r) = \lim_{p \to \infty} \lambda(r_p). \tag{6.13}$$

Let $i \in \{1, \ldots, m\}$. By (6.10) and (6.12),

$$r_p(i, a) + a \cdot I^{(p)} \to r(i, a) + a \cdot I \text{ as } p \to \infty$$

uniformly on $A(i)$. Together with (6.12) and (6.13) this implies that

$$\min_{a \in A(i)} [r(i, a) + a \cdot I] = \lim_{p \to \infty} \min_{a \in A(i)} [r_p(i, a) + a \cdot I^{(p)}]$$

$$= \lim_{p \to \infty} [I_i^{(p)} + \lambda(r_p)] = I_i + \lambda(r).$$

Lemma (6.2) is proved.

6.3 Proof of Theorem 6.1

Assume that the theorem does not hold. Then for each natural number p there exist $r_p \in C_{reg}(G)$ satisfying

$$\|r_p - \bar{r}\| \le p^{-1} \tag{6.14}$$

and an (r_p)-good pair $(x^{(p,0)}, \sigma^{(p)})$ where

$$\sigma^{(p)} = \{M^{(p,k)}\}_{k=0}^{\infty} \subset \mathcal{M}, \ x^{(p,0)} \in A \tag{6.15}$$

and

$$x^{(p,k+1)} = (M^{(p,k)})^T x^{(p,k)}, \ k = 0, 1, \ldots \tag{6.16}$$

such that

$$\limsup_{k \to \infty}(\|x^{(p,k)} - z(M^{(\bar{r})})\| + |M^{(p,k)} - M^{(\bar{r})}|_{\infty}) \ge \epsilon. \tag{6.17}$$

Let p be a natural number. In view of (6.17), there exists

$$(y^{(p)}, Q^{(p)}) \in \Omega(x^{(p,0)}, \sigma^{(p)}) \tag{6.18}$$

such that

$$\|y^{(p)} - z(M^{(\bar{r})})\| + |Q^{(p)} - M^{(\bar{r})}|_{\infty} \ge \epsilon/2. \tag{6.19}$$

By Propositions 3.7, 3.8 and (6.18) there exist a sequence

$$\{(y^{(p,k)}, Q^{(p,k)})\}_{k=-\infty}^{\infty} \subset \Omega(x^{(p,0)}, \sigma^{(p)}) \tag{6.20}$$

such that

$$y^{(p,0)} = y^{(p)}, \ Q^{(p,0)}) = Q^{(p)}, \tag{6.21}$$

for each integer k,

$$y^{(p,k+1)} = (Q^{(p,k)})^T y^{(p,k)} \tag{6.22}$$

and that for all integers k,

$$\sum_{i=1}^{m} y_i^{(p,k)} \theta^{(r_p)}(i, Q_i^{(p,k)}) = 0. \tag{6.23}$$

In view of (6.14),

$$\lim_{p\to\infty} \lambda(r_p) = \lambda(\bar{r}). \tag{6.24}$$

In view of (P5), extracting subsequences and re-indexing we may assume without loss of generality that there exists

$$I = \lim_{p\to\infty} I^{(r_p)}. \tag{6.25}$$

By Lemma 6.2, we may assume without loss of generality that

$$I = I^{(\bar{r})}. \tag{6.26}$$

It follows from (6.5), (6.14) and (6.24)–(6.26) that

$$\theta^{(r_p)}(i, a) \to \theta^{(\bar{r})}(i, a) \text{ as } p \to \infty \tag{6.27}$$

$$\text{uniformly for } (i, a) \in G.$$

Extracting subsequences, using a diagonalization process and re-indexing if necessary we may assume without loss of generality that for each integer k there exists

$$\bar{y}^{(k)} = \sum_{p\to\infty} y^{(p,k)}, \quad \bar{Q}^{(k)} = \lim_{p\to\infty} Q^{(p,k)}. \tag{6.28}$$

In view of (6.22) and (6.28), for all integers k,

$$\bar{y}^{(k+1)} = \lim_{p\to\infty} y^{(p,k+1)} = \lim_{p\to\infty} ((Q^{(p,k)})^T y^{(p,k)}) = (\bar{Q}^{(k)})^T \bar{y}^{(k)}. \tag{6.29}$$

Equations (6.19), (6.21) and (6.28) imply that

$$(\bar{y}^{(0)}, \bar{Q}^{(0)}) = \lim_{p\to\infty} (y^{(p)}, \bar{Q}^{(p)})$$

and

$$\|y^{(0)} - z(M^{(\bar{r})})\| + |Q^{(0)} - M^{(\bar{r})}|_\infty \geq \epsilon/4. \tag{6.30}$$

By (6.23), (6.27) and (6.28), for all integers k,

$$0 = \lim_{p\to\infty} \sum_{i=1}^{m} y_i^{(p,k)} \theta^{(r_p)}(i, Q_i^{(p,k)})$$

$$= \sum_{i=1}^{m} (\lim_{p \to \infty} y_i^{(p,k)} \lim_{p \to \infty} \theta^{(r_p)}(i, Q_i^{(p,k)}))$$

$$= \sum_{i=1}^{m} \bar{y}_i^{(k)} \theta^{(\bar{r})}(i, \bar{Q}_i^{(k)}). \tag{6.31}$$

Since the matrix $M^{(\bar{r})}$ is $\tilde{l}$-positive we have

$$((M^{(\bar{r})})^T)^k x \to z(M^{(\bar{r})}) \text{ as } k \to \infty \text{ uniformly on } A. \tag{6.32}$$

Let k be an integer and define

$$E_k = \{i \in \{1, \ldots, m\} : \bar{y}_i^{(k)} > 0\}. \tag{6.33}$$

Let $i \in E_k$. By (6.31) and (6.33),

$$\theta^{\bar{r}}(i, \bar{Q}_i^{(k)}) = 0.$$

Together with (6.5) and (6.9) this implies that

$$\bar{Q}_i^{(k)} = a^{(\bar{r},i)}, \ i \in E_k. \tag{6.34}$$

By (6.29), (6.33) and (6.34), for $j = 1, \ldots, m$,

$$\bar{y}_j^{(k+1)} = \sum_{i=1}^{m} \bar{y}_i^{(k)} \bar{Q}_{i,j}^{(k)} = \sum_{i \in E_k} \bar{y}_i^{(k)} \bar{Q}_{i,j}^{(k)} = \sum_{i \in E_k} \bar{y}_i^{(k)} a_j^{(\bar{r},i)}$$

$$= \sum_{i=1}^{m} \bar{y}_i^{(k)} a_j^{(\bar{r},i)} = ((M^{(\bar{r})})^T \bar{y}^{(k)})_j$$

and

$$\bar{y}^{(k+1)} = (M^{(\bar{r})})^T \bar{y}^{(k)} \text{ for all integers } k.$$

Together with (6.32) this implies that

$$\bar{y}^{(k)} = z(M^{(\bar{r})}) \text{ for all integers } k.$$

Combined with (6.33) and (6.34) this implies that

$$\bar{Q}^{(k)} = M^{(\bar{r})} \text{ for all integers } k.$$

This contradicts (6.50). The contradiction we have reached proves Theorem 6.1.

6.4 The Weak Turnpike Property

We continue to use the notation and definitions introduced in Sects. 2.1 2.3, 3.1, 3.2 and 6.1.

Assume that $\bar{r} \in C_{reg}(G)$, $\tilde{l}$ is a natural number, $M^{(\bar{r})}$ is $\tilde{l}$-positive and that

$$\text{if } I \in I(\bar{r}), \ i \in \{1, \ldots, m\}, \ b \in A(i),$$

$$I_i + \lambda(\bar{r}) = \bar{r}(i, b) + b \cdot I^{(\bar{r})}, \text{ then } b = a^{(\bar{r}, i)}. \tag{6.35}$$

We prove the following result which shows the stability of the weak turnpike property under small perturbations of the cost functions.

Theorem 6.3 *Let $\epsilon, c_0 > 0$. Then there exist a natural number L_1 and $\delta > 0$ such that for each $r \in C_{reg}(G)$ satisfying*

$$\|r - \bar{r}\| \leq \delta,$$

each $x^{(0)} \in A$, each Markov policy $\sigma = \{M^{(k)}\}_{k=0}^{\infty} \subset \mathcal{M}$ with

$$x^{(k+1)} = (M^{(k)})^T x^{(k)}, \ k = 0, 1, \ldots$$

and each natural number $N \geq L_1$ such that

$$C_N^r(x^{(0)}, \sigma) \leq \lambda(r)N + c_0$$

the inequality

$$Card(\{k \in \{0, \ldots, N-1\} : \ \|x^{(k)} - z(M^{(\bar{r})})\| + |M^{(k)} - M^{(\bar{r})}|_\infty > \epsilon\}) \leq L_1$$

holds.

6.5 Auxiliary Results for Theorem 6.3

Lemma 6.4 *Let $\epsilon > 0$. Then there exist $\delta > 0$ and a natural number N_1 such that for each $r \in C_{reg}(G)$ satisfying*

$$\|r - \bar{r}\| \leq \delta,$$

each $x^{(0)} \in A$ and each Markov policy $\{M^{(k)}\}_{k=0}^{\infty} \subset \mathcal{M}$,

$$x^{(k+1)} = (M^{(k)})^T x^{(k)}, \ k = 0, 1, \ldots$$

and a natural number $N > N_1$ such that

$$\sum_{k=0}^{N-1} \sum_{i=1}^{m} x_i^{(k)} \theta^{(r)}(i, M_i^{(k)}) \leq \delta$$

the following inequalities hold:

$$\|x^{(k)} - z(M^{(\bar{r})})\| \leq \epsilon, \ k = N_1, \dots, N-1,$$

$$|M^{(k)} - M^{(\bar{r})}|_\infty \leq \epsilon, \ k = N_1, \dots, N-1.$$

Proof Assume that the lemma does not hold. Then for each natural number p there exists $r_p \in C_{reg}(G)$ satisfying

$$\|r_p - \bar{r}\| \leq p^{-1}, \tag{6.36}$$

$$x^{(p,0)} \in A, \ \{M^{(p,k)}\}_{k=0}^{\infty} \subset \mathcal{M}$$

with

$$x^{(p,k+1)} = (M^{(p,k)})^T x^{(p,k)}, \ k = 0, 1, \dots, \tag{6.37}$$

a natural number $N_p > p$ such that

$$\sum_{k=0}^{N_p-1} \sum_{i=1}^{m} x_i^{(p,k)} \theta^{(r_p)}(i, M_i^{(p,k)}) \leq p^{-1} \tag{6.38}$$

and an integer

$$s_p \in [p, N_p - 1] \tag{6.39}$$

such that

$$\|x^{(p,s_p)} - z(M^{(\bar{r})})\| + |M^{(p,s_p)} - M^{(\bar{r})}|_\infty \geq \epsilon. \tag{6.40}$$

By (6.36),

$$\lim_{p \to \infty} \lambda(r_p) = \lambda(\bar{r}). \tag{6.41}$$

By (P5) and (6.36), the sequence $\{I^{(r_p)}\}_{p=0}^{\infty}$ is bounded. Extracting subsequences and re-indexing we may assume without loss of generality that

$$I = \lim_{p \to \infty} I^{(r_p)}. \tag{6.42}$$

In view of Lemma 6.2 and (6.42), we may assume without loss of generality that

$$I = I^{(\bar{r})}. \tag{6.43}$$

It follows from (6.5), (6.36), (6.41) and (6.42) that

$$\theta^{(r_p)}(i, a) \to \theta^{(\bar{r})}(i, a) \text{ as } p \to \infty \text{ uniformly on } G. \tag{6.44}$$

Let p be a natural number. Set

$$y^{(p,k)} = x^{(p,k+s_p+1)} \text{ for all integers } k \geq -s_p - 1, \tag{6.45}$$

$$Q^{(p,k)} = M^{(p,k+s_p+1)} \text{ for all integers } k \geq -s_p - 1. \tag{6.46}$$

By (6.37), (6.45) and (6.46), for all integers $k \geq -s_p - 1$,

$$y^{(p,k+1)} = (M^{(p,k+s_p+1)})^T x^{(p,k+s_p+1)} = (Q^{(p,k)})^T y^{(p,k)}. \tag{6.47}$$

In view of (6.38), (6.45) and (6.46), for all integers $k = -1 - s_p, \ldots, N_p - 2 - s_p$,

$$\sum_{i=1}^{m} y_i^{(p,k)} \theta^{(r_p)}(i, Q_i^{(p,k)}) = \sum_{i=1}^{m} x_i^{(p,k+s_p+1)} \theta^{(r_p)}(i, M_i^{(p,k+s_p+1)}) \leq p^{-1}. \tag{6.48}$$

Equations (6.40), (6.45) and (6.46) imply that

$$\|y^{(p,-1)} - z(M^{(\bar{r})})\| + |Q^{(p,-1)} - M^{(\bar{r})}|_\infty \geq \epsilon. \tag{6.49}$$

Extracting subsequences, using a diagonalization process and re-indexing we may assume without loss of generality that for every integer $k \leq 0$ there exist

$$y^{(k)} = \lim_{p \to \infty} y^{(p,k)}, \quad Q^{(k)} = \lim_{p \to \infty} Q^{(p,k)}. \tag{6.50}$$

By (6.47) and (6.50), for each integer $k < 0$.

$$y^{(k+1)} = (Q^{(k)})^T y^{(k)}. \tag{6.51}$$

It follows from (6.39), (6.48) and (6.50) that for each integer $k < 0$,

$$\sum_{i=1}^{m} y_i^{(k)} \theta^{(\bar{r})}(i, Q_i^{(k)}) = 0. \tag{6.52}$$

In view of (6.49) and (6.50),

$$\|y^{(-1)} - z(M^{(\bar{r})})\| + |Q^{(-1)} - M^{(\bar{r})}|_\infty \geq \epsilon/2. \tag{6.53}$$

Lemma 4.2 implies that $y^{(k)} = z(M^{(\bar{r})})$ for all integers $k \leq 0$ and $M^{(k)} = M^{(\bar{r})}$ for all integers $k < 0$. This contradicts (6.53). The contradiction we have reached proves Lemma 6.4.

6.6   Proof of Theorem 6.3

By Lemma 6.4, there exist $\delta > 0$ and a natural number N_1 such that the following property holds:

(P6)   for each $r \in C_{reg}(G)$ satisfying

$$\|r - \bar{r}\| \leq \delta,$$

each $x^{(0)} \in A$, each Markov policy $\{M^{(k)}\}_{k=0}^\infty \subset \mathcal{M}$ with

$$x^{(k+1)} = (M^{(k)})^T x^{(k)}, \quad k = 0, 1, \ldots$$

and each natural number $N > N_1$ such that

$$\sum_{k=0}^{N-1} \sum_{i=1}^{m} x_i^{(k)} \theta^{(r)}(i, M_i^{(k)}) \leq \delta$$

the following inequalities hold:

$$\|x^{(k)} - z(M^{(\bar{r})})\| \leq \epsilon/4, \quad k = N_1, \ldots, N - 1,$$

$$|M^{(k)} - M^{(\bar{r})}|_\infty \leq \epsilon/4, \quad k = N_1, \ldots, N - 1.$$

Property (P5) implies that there exists $c_1 > 0$ such that the following property holds:

$|I|_\infty \leq c_1$ for each $r \in C_{reg}(G)$ satisfying $\|r - \bar{r}\| \leq 1$ and each $I \in I(r)$.

Choose a natural number

$$L_1 \geq (2N_1 + 7)(\delta^{-1}(c_0 + 2c_1) + 1). \tag{6.54}$$

Assume that $r \in C_{reg}(G)$,

$$\|r - \bar{r}\| \leq \delta,$$

$$x^{(0)} \in A, \ \{M^{(k)}\}_{k=0}^{\infty} \subset \mathcal{M},$$

$$x^{(k+1)} = (M^{(k)})^T x^{(k)}, \ k = 0, 1, \ldots, \tag{6.55}$$

a natural number $N \geq L_1$ and that

$$C_N^r(x^{(0)}, \sigma) \leq \lambda(r)N + c_0. \tag{6.56}$$

Proposition 3.5 and Eqs. (6.55), (6.56) imply that

$$c_0 + \lambda(r)N \geq C_N^r(x^{(0)}, \sigma)$$

$$= \lambda(r)N + \sum_{k=0}^{N-1}\sum_{i=1}^{m} x_i^{(k)}\theta^{(r)}(i, M_i^{(k)}) + x^{(0)} \cdot I - x^{(N)} \cdot I.$$

By the inequality above, (6.55),

$$\sum_{k=0}^{N-1}\sum_{i=1}^{m} x_i^{(k)}\theta^{(r)}(i, M_i^{(k)}) \leq c_0 + 2c_1. \tag{6.57}$$

Set $t_0 = 0$. By induction we will define a finite sequence of a strictly increasing integers $t_0, \ldots, t_s$.

Assume that $j \geq 0$ is an integer and that we have already defined integers $t_0, \ldots, t_j$. If $t_j = N$, then the construction is completed. Assume that $t_j < N$. If

$$\sum_{k=t_j}^{N-1}\sum_{i=1}^{m} x_i^{(k)}\theta^{(r)}(i, M_i^{(k)}) \leq \delta,$$

then we set

$$t_{j+1} = N \tag{6.58}$$

and the construction is completed.

Assume that

$$\sum_{k=t_j}^{N-1}\sum_{i=1}^{m} x_i^{(k)}\theta^{(r)}(i, M_i^{(k)}) > \delta. \tag{6.59}$$

Then there exists a unique integer $t_{j+1} > t_j$ such that $t_{j+1} \le N$,

$$\sum_{k=t_j}^{t_{j+1}-1} \sum_{i=1}^{m} x_i^{(k)} \theta^{(r)}(i, M_i^{(k)}) > \delta \tag{6.60}$$

and for any integer p satisfying

$$t_j < p < t_{j+1}, \tag{6.61}$$

we have

$$\sum_{k=t_j}^{p-1} \sum_{i=1}^{m} x_i^{(k)} \theta^{(r)}(i, M_i^{(k)}) \le \delta. \tag{6.62}$$

Thus by induction we have constructed a strictly increasing sequence of integers $t_0, t_1, \ldots$ which is finite in view of (6.60). Let t_s be its last element. Clearly, $t_s = N$ and $s \ge 1$. By (6.57) and (6.60),

$$c_0 + 2c_1$$

$$\ge \sum \{ \sum_{k=t_j}^{t_{j+1}-1} \sum_{i=1}^{m} x_i^{(k)} \theta^{(r)}(i, M_i^{(k)}) : \text{ an integer } j$$

$$\text{satisfies } 0 \le j \le s - 2 \} \ge \delta(s - 1)$$

and

$$s \le \delta^{-1}(c_0 + 2c_1) + 1. \tag{6.63}$$

Define

$$E = \{ j \in \{0, \ldots, s - 1\} : t_{j+1} - t_j \ge N_1 + 4 \}. \tag{6.64}$$

Let

$$j \in E.$$

By (6.64),

$$t_{j+1} - t_j > N_1 + 4. \tag{6.65}$$

and by the definition of t_{j+1}, (6.58) and (6.62),

$$\sum_{k=t_j}^{t_{j+1}-2} \sum_{i=1}^{m} x_i^{(k)} \theta^{(r)}(i, M_i^{(k)}) \le \delta. \tag{6.66}$$

By (6.55), (6.65) and (6.66) and property (P6) applied to the initial distribution $x^{(t_j)}$ and to the Markov policy $\{M^{(t_j+k)}\}_{k=0}^{\infty}$, for all integers $k = t_j + N_1, \ldots, t_{j+1} - 2$,

$$\|x^{(k)} - z(M^{(\bar{r})})\| \le \epsilon/4, \quad |M^{(k)} - M^{(\bar{r})}|_{\infty} \le \epsilon/4.$$

This implies that

$$F := \{k \in \{0, \ldots, N-1\} : \|x^{(k)} - z(M^{(\bar{r})})\| + |M^{(k)} - M^{(\bar{r})}|_{\infty} > \epsilon\}$$

$$\subset \cup\{\{t_j, \ldots, t_{j+1}\} : j \in \{0, \ldots, s-1\} \setminus E\}$$

$$\cup \{\{t_j, \ldots, t_j + N_1 - 1\} \cup \{t_{j+1} - 1, t_{j+1}\} : j \in E\}. \tag{6.67}$$

By (6.54), (6.63), (6.64) and (6.67)

$$\mathrm{Card}(F) \le s(N_1 + 5) + s(N_1 + 2)$$

$$\le (2N_1 + 7)(\delta^{-1}(c_0 + 2c_1) + 1) \le L_1.$$

Theorem 6.3 is proved.

Chapter 7
Controllability Properties

In this chapter we discuss controllability of discrete-time deterministic control systems which are useful in the turnpike theory. We introduce a notion of an $\mathcal{M}$-positive matrix and show that under certain condition for a generic cost function the corresponding MDP has the unique minimizing Markov actions which form a $\mathcal{M}$-positive matrix. In this case our controllability results are true and the strong turnpike property holds and is stable. This will be shown in Chap. 8.

7.1 Linear Systems in Abstract Spaces

Assume that $(Z, \langle \cdot, \cdot \rangle)$ is a Hilbert space equipped with the inner product $\langle \cdot, \cdot \rangle$ which induces the norm $\|z\| = \langle z, z \rangle^{1/2}$, $z \in Z$. Denote by $\mathcal{L}(Z, Z)$ the space of all bounded linear continuous mappings $L : Z \to Z$. For each $L \in \mathcal{L}(Z, Z)$ set

$$\|L\| = \sup\{\|L(z)\|\|z\|^{-1} : z \in Z \setminus \{0\}\}.$$

We begin with the following controllability result.

Proposition 7.1 *Assume that $L \in \mathcal{L}(Z, Z)$, $x \in Z$, $\|x\| > 0$,*

$$L(x) = x, \tag{7.1}$$

$\epsilon \in (0, 1)$, $\xi_1, \xi_2 \in Z$,

$$\|\xi_i - x\| \leq 2^{-1}\epsilon \min\{1, \|x\|\}(1 + \|L\|)^{-1}, \ i = 1, 2. \tag{7.2}$$

Then there exists $L_1 \in \mathcal{L}(Z, Z)$ such that

$$\|L - L_1\| \leq \epsilon, \ L_1(\xi_1) = \xi_2.$$

© The Author(s), under exclusive license to Springer Nature Switzerland AG 2025
A. J. Zaslavski, *Turnpike Phenomenon for Markov Decision Processes*,
SpringerBriefs in Mathematics, https://doi.org/10.1007/978-3-032-00854-1_7

Proof Set

$$Z_0 = \{\lambda \xi_1 : \lambda \in R^1\}. \tag{7.3}$$

By (7.2),

$$\|\xi_1\| = \|x + \xi_1 - x\| \geq \|x\| - \|\xi_1 - x\| \geq 2^{-1}\|x\|. \tag{7.4}$$

Let $P : Z \to Z_0$ be a projection operator. We have

$$\|P(u)\| \leq \|u\|, \ u \in Z. \tag{7.5}$$

Consider a linear operator $Q : Z_0 \to Z$ such that

$$Q(\xi_1) = \xi_2 - x - L(\xi_1 - x) \tag{7.6}$$

and set

$$L_1 = L + QP. \tag{7.7}$$

By (7.2), (7.4) and (7.6),

$$\|Q\| = \|\xi_2 - x\|\|\xi_1\|^{-1} \leq \|L\|\|\xi_1 - x\|\|\xi_1\|^{-1} \leq \epsilon \min\{1, \|x\|\}\|x\| \leq \epsilon. \tag{7.8}$$

In view of (7.5) and (7.8),

$$\|L_1 - L\| = \|QP\| \leq \|Q\|\|P\| \leq \epsilon.$$

Equations (7.1), (7.3), (7.6) and (7.7) imply that

$$L_1(\xi_1) = L(\xi_1) + (L_1 - L)(\xi_1)$$

$$= L(x) + L_1(\xi_1 - x) + QP(\xi_1) = L(x) + L(\xi_1 - x) + Q(\xi_1)$$

$$= x + L(\xi_1 - x) + \xi_2 - x - L(\xi_1 - x) = \xi_2.$$

Proposition 7.1 is proved.

For each $L \in \mathcal{L}(Z, Z)$ denote by L^* its dual operator.

Using Proposition 7.1 we prove our next controllability result which can be applied to the deterministic control system associated with our MDP.

Proposition 7.2 *Let* $x \in Z, \|x\| > 0, l \in Z, \|l\| = 1, L \in \mathcal{L}(Z, Z),$

$$L(x) = x, \tag{7.9}$$

$$L^*(l) = l \circ L = l, \tag{7.10}$$

$\epsilon \in (0, 1),$

$$\delta = 2^{-1}\epsilon \|x\|(1 + \|L\|)^{-1}, \tag{7.11}$$

$\xi_1, \xi_2 \in Z,$

$$\|\xi_i - x\| \leq \delta, \ i = 1, 2, \tag{7.12}$$

$$\langle l, \xi_i \rangle = \langle l, x \rangle, \ i = 1, 2. \tag{7.13}$$

Then there exists $L_1 \in \mathcal{L}(Z, Z)$ such that

$$\|L_1 - L\| \leq \epsilon,$$

$$l \circ L_1 = L_1^*(l) = l, \ L_1(\xi_1) = \xi_2.$$

Proof Set

$$Z_0 = \{z \in Z : \ \langle l, z \rangle = 0\}. \tag{7.14}$$

Clearly, Z_0 is a closed linear subspace of Z and

$$Z_0 + \{\lambda l : \ \lambda \in R^1\} = Z.$$

Consider the Hilbert subspace $Z_0 \subset Z$ with the induced inner product. By (7.10) and (7.14), for each $z \in Z_0$,

$$\langle l, L(z) \rangle = \langle L^*(l), z_0 \rangle = \langle l, z_0 \rangle = 0,$$

$$L(z) \in Z_0, \ L(Z_0) \subset Z_0. \tag{7.15}$$

Consider the projection operator $P : Z \to Z_0$. Assume that

$$\xi_1, \xi_2 \in Z$$

and (7.12), (7.13) hold. Proposition 7.1 and (7.9), (7.11), (7.12) imply that there exists $\tilde{L} : Z \to Z$ such that

$$\|\tilde{L}\| \leq \epsilon, \tag{7.16}$$

$$(L + \tilde{L})(\xi_1) = \xi_2. \tag{7.17}$$

Set

$$L_1 = L + P\tilde{L}. \tag{7.18}$$

In view of (7.16) and (7.18),

$$\|L_1 - L\| \le \|P\tilde{L}\| \le \|\tilde{L}\| \le \epsilon.$$

By (7.10) and (7.18), for each $z \in Z$,

$$\langle l, L_1(z) \rangle = \langle l, L(z) \rangle + \langle l, P\tilde{L}(z) \rangle = \langle l, z \rangle.$$

It follows from (7.17) and (7.18) that

$$L_1(\xi_1) = (L + P\tilde{L})(\xi_1)$$

$$= (L + \tilde{L})(\xi_1) + P\tilde{L}(\xi_1) - \tilde{L}(\xi_1) = \xi_2 + P\tilde{L}(\xi_1) - \tilde{L}(\xi_1). \tag{7.19}$$

By (7.10), (7.13), (7.14) and (7.17),

$$\langle l, \tilde{L}(\xi_1) \rangle = \langle l, \xi_2 - L(\xi_1) \rangle = \langle l, \xi_2 \rangle - \langle l, L(\xi_1) \rangle = \langle l, \xi_2 \rangle - \langle l, \xi_1 \rangle = 0,$$

$$\langle \tilde{l}, \tilde{L}(\xi_1) \rangle = 0, \ \tilde{L}(\xi_1) \in Z_0,$$

$$P\tilde{L}(\xi_1) = \tilde{L}(\xi_1).$$

This implies that $L_1(\xi_1) = \xi_2$. Proposition 7.2 is proved.

7.2 Auxiliary Results

Let $(Y, \| \cdot \|)$ be a normed linear space. For each $y \in Y$ and $r > 0$ set

$$B(y, r) = \{z \in Y : \ \|z - y\| \le r\}.$$

Lemma 7.3 ([83]) *Let $(Y, \| \cdot \|)$ be a normed linear space and let $r > 0$ be given. Assume that C is a closed and convex subset of Y such that for all $y \in B(0, r)$,*

$$\inf_{x \in C} \|y - x\| \le r. \tag{7.20}$$

Then $0 \in C$.

Proof Assume that $0 \notin C$. Then by the separation theorem there exists a bounded linear functional $l \in Y^*$ (the dual space) such that $\|l\|_* = 1$ and

$$p = \inf\{l(x) : \ x \in C\} > 0.$$

There is $y_0 \in B(0, r)$ such that $l(-y_0) > r - p/2$. By (7.20), there is $x_0 \in C$ such that $\|y_0 - x_0\| < r + p/2$. Now we have

$$p \le l(x_0) = l(y_0) + l(x_0 - y_0) < -r + p/2 + \|x_0 - y_0\|$$

$$< -r + p/2 + r + p/2 = p.$$

Since we have reached a contradiction, we conclude that the origin does belong to C. Lemma 7.3 is proved.

Denote by $\mathcal{L}(Y, Y)$ the set of all continuous bounded linear operators $T : Y \to Y$ with

$$\|T\| = \sup\{\|T(y)\| : y \in Y, \ \|y\| \le 1\}.$$

Proposition 7.4 *Assume that $\xi \in Y$,*

$$\mathcal{L}_\xi = \{T \in \mathcal{L}(Y, Y) : \ T(\xi) = \xi\},$$

$T_0 \in \mathcal{L}_\xi$, $C \subset \mathcal{L}_\xi$ is a nonempty closed convex set, $r > 0$ and that for each $T \in \mathcal{L}_\xi$ satisfying $\|T - T_0\| \le r$,

$$\inf\{\|S - T\| : \ S \in C\| \le r.$$

Then $T_0 \in C$.

Proof Set

$$\tilde{\mathcal{L}} = \mathcal{L}_\xi - T_0 = \{T \in \mathcal{L}(Y, Y) : \ T(\xi) = 0\}.$$

Clearly, $\tilde{\mathcal{L}}$ is a closed linear subspace of $\mathcal{L}(Y, Y)$ and $C - T_0 \subset \tilde{\mathcal{L}}$ is a closed convex set.

Assume that $T \in \tilde{\mathcal{L}}$,

$$\|T\| \le r.$$

Then

$$T + L_0 \in \mathcal{L}_\xi,$$

$$r \ge \inf\{\|S - T - T_0\| : \ S \in C\|\} = \inf\{\|Q - T\| : \ Q \in C - T_0\|\}.$$

Together with Lemma 7.3 applied with $Y = \tilde{\mathcal{L}}$ and the set $C - T_0$ this implies

$$0 \in C - T_0, \quad T_0 \in C.$$

Proposition 7.4 is proved.

7.3   Finite-Dimensional Spaces

We use the notation of Chap. 2 and show how the results of the previous sections can be applied to the deterministic optimal control problem associated with our MDP.

Proposition 7.5 *Assume that $x \in R^m$, $x >> 0$, $\sum_{i=1}^{m} x_i = 1$, $M \in \mathcal{M}$, all entries of M are positive, $M^T x = x$ and there is $\Delta > 0$ such that each matrix $\tilde{M} = (\tilde{M}_{i,j})$, $i, j = 1, \ldots, m$ satisfying*

$$|M_{i,j} - \tilde{M}_{i,j}| \leq \Delta, \quad i, j = 1, \ldots, m,$$

$$\tilde{M}(e) = M(1, \ldots, 1) = 1$$

belongs to $\mathcal{M}$. Then for each $\epsilon > 0$ there exists $\delta > 0$ such that for each $\xi_1, \xi_2 \in A$ satisfying $\|\xi_i - x\| \leq \delta$, $i = 1, 2$ there exists $S \in \mathcal{M}$ which satisfies

$$|S_{i,j} - M_{i,j}| \leq \epsilon, \quad i, j = 1, \ldots, m,$$

$$S(\xi_1) = \xi_2.$$

Proposition 7.5 follows from Proposition 7.2.

We identify a $m \times m$ matrix M with the linear operator in R^m equipped with the Euclidean norm, induced by this matrix, and denote by $\|M\|$ the norm of this operator.

Proposition 7.6 *Assume that $M \in \mathcal{M}$, $M_i \in A(i)$, $i = 1, \ldots, m$, all the entries of M are positive, there is $\Delta > 0$ such that each matrix $S = (S_{i,j})$, $i, j = 1, \ldots, m$ satisfying*

$$\|S - M\| \leq \Delta, \quad S(e) = e$$

belongs to $\mathcal{M}$, $r > 0$, C is a nonempty closed convex set of $m \times m$ matrices, for each $T \in C$, $T(e) = e$ and that for each $T \in \mathcal{M}$ satisfying $\|T - M\| \leq r$,

$$\inf\{\|S - T\| : \ S \in C\|\} \leq r.$$

Then $M \in C$.

Proposition 7.6 follows from Proposition 7.4.

7.4 A Subclass of MDPs

We use the notation and definitions of Sects. 2.1, 2.3 and 3.1.

Denote by $\mathfrak{M}$ the set of all $m \times m$ matrix such that for each $i \in \{1, \ldots, m\}$, $M_i \in A$.

A matrix $M \in \mathcal{M}$ is called $\mathcal{M}$-positive if all its entries are positive and there exists $\Delta_M > 0$ such that each $\tilde{M} \in \mathfrak{M}$ satisfying $\|\tilde{M} - M\| \leq \Delta_M$ belongs to $\mathcal{M}$.

In view of Propositions 7.5 and 7.6, the strong turnpike property and its stability are obtained for MDPs for which minimizing Markov actions are in fact lines of a $\mathcal{M}$-positive matrix.

In the sequel we use the following condition.

Condition (E) The set of all $\mathcal{M}$-positive matrices is an everywhere dense subset of $\mathcal{M}$ equipped with the natural topology induced by the Euclidean metric.

Proposition 7.7 *Assume that for each* $i \in \{1, \ldots, m\}$, $A(i)$ *is convex and that* $\tilde{M} \in \mathcal{M}$ *is* $\mathcal{M}$-*positive. Then condition (E) holds.*

Proof Clearly, for each $\alpha \in (0, 1)$ and each $M \in \mathcal{M}$,

$$\alpha M + (1 - \alpha)\tilde{M} \in \mathcal{M}$$

is $\mathcal{M}$-positive and condition (E) holds. Proposition 7.7 is true.

We prove the following result which shows that for a generic cost function equation (2.1) has a unique solution which forms an $\mathcal{M}$-positive matrix.

Theorem 7.8 *Let* A *be either* $C_{reg}(G)$ *or* $C_{conv}(G)$ *(if* $A(i)$ *is convex for all* $i = 1, \ldots, m$). *Assume that condition (E) holds. Then there exists a set* $\mathcal{F} \subset A$ *which is a countable intersection of open everywhere dense subsets of* A *such that for each* $r \in \mathcal{F}$ *the following property holds:*

There exist actions $a^{(r,i)} \in A(i)$, $i = 1, \ldots, m$ *such that the matrix* $M \in \mathcal{M}$ *with* $M_i = a^{(r,i)}$, $i = 1, \ldots, m$ *is* $\mathcal{M}$-*positive. If* $I \in R^m$, $a^i \in A(i)$, $i = 1, \ldots, m$ *satisfy*

$$I_i + \lambda(r) = \min_{a \in A(i)} [r(i, a) + a \cdot I] = r(i, a^i) + a^i \cdot I, \ i = 1, \ldots, m,$$

then $a^i = a^{(r,i)}$, $i = 1, \ldots, m$.

Thus under condition (E) for a generic cost function the corresponding MDP has the unique minimizing Markov actions which form a $\mathcal{M}$-positive matrix. In this case our controllability results (Propositions 7.5 and 7.6) are true and the strong turnpike property holds and is stable. This will be shown in Chap. 8. Namely in Chap. 8 the strong turnpike property and its stability are established for a general optimal control problem and our MDP with the unique minimizing Markov actions forming a $\mathcal{M}$-positive matrix is its particular case.

For the proof of Theorem 7.8 we need the following auxiliary result which is proved analogously to Lemma 2.8.

Lemma 7.9 *Assume that condition (E) holds, $r \in C_{reg}(G)$ and $\epsilon > 0$. Then there exist $\bar{r} \in C_{reg}(G)$, $\bar{I} \in R^m$, $b^i \in A(i)$, $i = 1, \ldots, m$ and $\mathcal{M}$-positive $M \in \mathcal{M}$ such that*

$$\|\bar{r} - r\| \leq \epsilon,$$

$$\bar{I}_i + \lambda(\bar{r}) = \min_{a \in A(i)} [\bar{r}(i, a) + a \cdot \bar{I}] = \bar{r}(i, b^i) + b^i \cdot \bar{I}, \ i = 1, \ldots, m,$$

$$M_i = b^i, \ i = 1, \ldots, m.$$

Moreover if for all $i \in \{1, \ldots, m\}$, $A(i)$ is convex and $r(i, \cdot) : A(i) \to R^1$ is convex for all $i = 1, \ldots, m$, then $\bar{r}(i, \cdot) : A(i) \to R^1$ is convex for all $i = 1, \ldots, m$.

Proof of Theorem 7.8 Let $r \in \mathcal{A}$, $\gamma \in (0, 1)$. By Lemma 7.9 there exist $r_{\gamma,0} \in \mathcal{A}$, $I \in R^m$, $a^{(r,i)} \in A(i)$, $i = 1, \ldots, m$ such that

$$\|r_{\gamma,0} - r\| \leq \gamma/2, \tag{7.21}$$

$$I_i + \lambda(r_{\gamma,0}) = \min_{a \in A(i)} [r_{\gamma,0}(i, a) + a \cdot I] = r_{\gamma,0}(i, a^{(r,i)}) + a^{(r,i)} \cdot I, \ i = 1, \ldots, m \tag{7.22}$$

and the matrix $\tilde{M}$ with

$$\tilde{M}_i = a^{(r,i)}, \ i = 1, \ldots, m \tag{7.23}$$

is $\mathcal{M}$-positive. Define

$$r_\gamma(i, a) = r_{\gamma,0}(i, a) + \gamma \|a - a^{(r,i)}\|, \ (i, a) \in G. \tag{7.24}$$

By Lemma 2.9,

$$r_\gamma \in \mathcal{A}$$

and for each $i \in \{1, \ldots, m\}$,

$$I_i + \lambda(r_\gamma) = \min_{a \in A(i)} [r_\gamma(i, a) + a \cdot I] = r_\gamma(i, a^{(r,i)}) + a^{(r,i)} \cdot I \tag{7.25}$$

and the following property holds:

(a) if

$$\tilde{I} = (\tilde{I}_1, \ldots, \tilde{I}_m) \in R^m, \ b^i \in A(i), \ i = 1, \ldots, m,$$

$$\tilde{I}_i + \lambda(r_\gamma) = \min_{a \in A(i)} [r_\gamma(i, a) + a \cdot \tilde{I}] = r_\gamma(i, b^i) + b^i \cdot \tilde{I}, \ i = 1, \ldots, m,$$

then $b^i = a^{(r,i)}, i = 1, \ldots, m$.

Since the matrix $\tilde{M}$ is $\mathcal{M}$-positive there exists an open neighborhood $V(r, \gamma)$ of $\tilde{M}$ in $\mathfrak{M}$ such that

$$V(r, \gamma) \subset \mathcal{M}, \ \text{each } M \in V(r, \gamma) \text{ is } \mathcal{M} - \text{positive.} \tag{7.26}$$

In view of (a) and Proposition 2.11, there exists an open neighborhood $\mathcal{U}(r, \gamma)$ of r_γ in $\mathcal{A}$ such that the following properties hold:

(b) If

$$h \in \mathcal{U}(r, \gamma, p), \ I \in R^m, \ b^i \in A(i), \ i = 1, \ldots, m,$$

$$I_i + \lambda(h) = \min_{a \in A(i)} [h(i, a) + a \cdot I] = h(i, b^i) + b^i \cdot I, \ i = 1, \ldots, m,$$

$M \in \mathcal{M}$ satisfies $M_i = b^i, i = 1, \ldots, m$, then $M \in V(r, \gamma)$.

Define

$$\mathcal{F}_0 = \cup \{\mathcal{U}(r, \gamma) : \ r \in \mathcal{A}, \ \gamma \in (0, 1)\}. \tag{7.27}$$

Clearly, $\mathcal{F}_0$ is an open everywhere dense subset of $\mathcal{A}$.

Assume that

$$h \in \mathcal{F}_0, \tag{7.28}$$

$$I \in R^m, \ b^i \in A(i), \ i = 1, \ldots, m,$$

$$I_i + \lambda(h) = \min_{a \in A(i)} [h(i, a) + a \cdot I] = h(i, b^i) + b^i \cdot I, \ i = 1, \ldots, m. \tag{7.29}$$

By (7.27) and (7.28), there exist $r \in \mathcal{A}, \gamma \in (0, 1)$ such that

$$h \in \mathcal{U}(r, \gamma). \tag{7.30}$$

Let $M \in \mathcal{M}$ and $M_i = b^i, i = 1, \ldots, m$. Property (b) and (7.26), (7.27), (7.29) and (7.30) imply that $M \in V(r, \gamma)$ and M is $\mathcal{M}$-positive. Together with Theorem 2.7 this implies Theorem 7.8.

Chapter 8
Optimal Control Problems with Singleton-Turnpikes

In this chapter we study the structure of approximate solutions of a general optimal control system. The deterministic optimal control system associated with a MDP is its particular case if the unique minimizing Markov actions of this MDP form a $\mathcal{M}$-positive matrix. We show that the strong turnpike property holds and that it is stable under small perturbations of the parameters of the system. The strong turnpike properties for deterministic optimal control systems are well-known in the literature but they are related to states of the system. Here we study the situation when the turnpike phenomenon holds both for its states and its controls.

8.1 Preliminaries and Stability Results

For each pair of nonempty sets X, Y and each mapping $T : X \to 2^Y$ set

$$\mathrm{graph}(T) = \{(x, y) :\ x \in X,\ y \in T(x)\}.$$

Let (E, ρ_E) and (F, ρ_F) be compact metric spaces. For each $x \in E$, each $y \in F$, each $r > 0$, each $C \subset E$ and each $D \subset F$ set

$$\rho_E(x, C) = \inf\{\rho_E(x, z) :\ z \in C\},$$

$$\rho_F(y, D) = \inf\{\rho_F(y, z) :\ z \in D\},$$

$$B_E(x, r) = \{z \in E :\ \rho_E(x, z) \le r\},$$

$$B_F(y, r) = \{z \in F :\ \rho_F(y, z) \le r\}.$$

We suppose that the infimum over empty set if ∞ and that the sum over empty set is zero.

© The Author(s), under exclusive license to Springer Nature Switzerland AG 2025
A. J. Zaslavski, *Turnpike Phenomenon for Markov Decision Processes*,
SpringerBriefs in Mathematics, https://doi.org/10.1007/978-3-032-00854-1_8

The space $E \times F$ is equipped with the metric

$$\rho_{E \times F}((x_1, u_1), (x_2, u_2)) = \rho_E(x_1, x_2) + \rho_F(u_1, u_2), \ (x_i, u_i) \in E \times F, \ i = 1, 2$$

and the space $E \times F \times E$ is equipped with the metric

$$\rho_{E \times F \times E}((x_1, u_1, z_1), (x_2, u_2, z_2)) = \rho_E(x_1, x_2) + \rho_F(u_1, u_2) + \rho_E(z_1, z_2),$$

$$(x_i, u_i, z_i) \in E \times F \times E, \ i = 1, 2.$$

Denote by $\mathfrak{M}$ the set of all bounded functions

$$f : \{0, 1, \dots\} \times E \times F \to R^1.$$

For each $f \in \mathfrak{M}$ set

$$\|f\| = \sup\{|f(t, x, u)| : \ (t, x, u) \in \{0, 1, \dots\} \times E \times F\}. \tag{8.1}$$

We suppose that $\mathcal{A}$ is a nonempty subset of $\{0, 1, \dots, \} \times E, \mathcal{U} : \mathcal{A} \to 2^F$ is a point to set mapping with a graph

$$\mathcal{M} = \mathrm{graph}(\mathcal{U}) := \{(t, x, u) : \ (t, x) \in \mathcal{A}, \ u \in \mathcal{U}(t, x)\},$$

$G : \mathcal{M} \to E$ and $f \in \mathfrak{M}$.

Let $0 \le T_1 < T_2$ be integers. We denote by $X(T_1, T_2, \mathcal{M}, G)$ the set of all pairs of sequences $(\{x_t\}_{t=T_1}^{T_2}, \{u_t\}_{t=T_1}^{T_2-1})$ such that each $t \in \{T_1, \dots, T_2 - 1\}$,

$$x_t \in \mathcal{A}, \ u_t \in \mathcal{U}(t, x_t), \ x_{t+1} = G(t, x_t, u_t) \tag{8.2}$$

and which are called trajectory-control pairs. For simplicity we use the notation $X(T_1, T_2) = X(T_1, T_2, \mathcal{M}, G)$ is the pair $(\mathcal{M}, G)$ is understood.

Let $T_1 \ge 0$ be an integer. Denote by $X(T_1, \infty, \mathcal{M}, G)$ the set of all pairs of sequences $\{x_t\}_{t=T_1}^{\infty} \subset E, \{u_t\}_{t=T_1}^{\infty} \subset F$ such that for each integer $T_2 > T_1, (\{x_t\}_{t=T_1}^{T_2}, \{u_t\}_{t=T_1}^{T_2-1}) \in X(T_1, T_2, \mathcal{M}, G)$. Elements of $X(T_1, \infty, \mathcal{M}, G)$ are called trajectory-control pairs. For simplicity we use the notation $X(T_1, \infty) = X(T_1, \infty, \mathcal{M}, G)$ is the pair $(\mathcal{M}, G)$ is understood.

Let $0 \le T_1 < T_2$ be integers. A sequence $\{x_t\}_{t=T_1}^{T_2} \subset E$ ($\{x_t\}_{t=T_1}^{\infty} \subset E$ respectively) is called a trajectory if there exists a sequence $\{u_t\}_{t=T_1}^{T_2-1} \subset F$ ($\{u_t\}_{t=T_1}^{\infty} \subset F$ respectively) referred to as a control such that

$$(\{x_t\}_{t=T_1}^{T_2}, \{u_t\}_{t=T_1}^{T_2-1}) \in X(T_1, T_2)$$

$(\{x_t\}_{t=T_1}^{\infty}, \{u_t\}_{t=T_1}^{\infty}) \in X(T_1, \infty)$ respectively).

Let $f \in \mathfrak{M}$. For each pair of integers $T_2 > T_1 \geq 0$ and each pair of points $y, z \in E$ we consider the following problems:

$$\sum_{t=T_1}^{T_2-1} f(t, x_t, u_t) \to \min,$$

$$(\{x_t\}_{t=T_1}^{T_2}, \{u_t\}_{t=T_1}^{T_2-1}) \in X(T_1, T_2, \mathcal{M}, G)), \ x_{T_1} = y, \ x_{T_2} = z, \qquad (P_1)$$

$$\sum_{t=T_1}^{T_2-1} f(t, x_t, u_t) \to \min, \ (\{x_t\}_{t=T_1}^{T_2}, \{u_t\}_{t=T_1}^{T_2-1}) \in X(T_1, T_2, \mathcal{M}, G), \ x_{T_1} = y,$$
$$(P_2)$$

$$\sum_{t=T_1}^{T_2-1} f(t, x_t, u_t) \to \min, \ (\{x_t\}_{t=T_1}^{T_2}, \{u_t\}_{t=T_1}^{T_2-1}) \in X(T_1, T_2, \mathcal{M}, G). \qquad (P_3)$$

For each pair of integers $T_2 > T_1 \geq 0$ and each pair of points $y, z \in E$ we define

$$U_{\mathcal{M},G}^f(T_1, T_2, y, z) = \inf\{ \sum_{t=T_1}^{T_2-1} f(t, x_t, u_t) :$$

$$(\{x_t\}_{t=T_1}^{T_2}, \{u_t\}_{t=T_1}^{T_2-1}) \in X(T_1, T_2, \mathcal{M}, G), \ x_{T_1} = y, \ x_{T_2} = z\},$$

$$U_{\mathcal{M},G}^f(T_1, T_2, z) = \inf\{U_{\mathcal{M},G}^f(T_1, T_2, z, h) : \ h \in E\},$$

$$\widehat{U}_{\mathcal{M},G}^f(T_1, T_2, z) = \inf\{U_{\mathcal{M},G}^f(T_1, T_2, y, z) : y \in E\},$$

$$U_{\mathcal{M},G}^f(T_1, T_2) = \inf\{U_{\mathcal{M},G}^f(T_1, T_2, y, z) : y, z \in E\}. \qquad (8.3)$$

We suppose that $f_0 : E \times F \to R^1$ is a bounded function,

$$f_*(t, x, u) = f_0(x, u), \ (t, x, u) \in \{0, 1, \ldots\} \times E \times F, \qquad (8.4)$$

$\mathcal{A}_0 \subset E$ is a nonempty set, $\mathcal{U}_0 : \mathcal{A}_0 \to 2^F$ is a point to set mapping,

$$\mathcal{A}_* = \{0, 1, \ldots\} \times \mathcal{A}_0, \ \mathcal{U}_*(t, x) = \mathcal{U}_0(x), \ x \in \mathcal{A}_0, \ t \in \{0, 1, \ldots\}, \qquad (8.5)$$

$$\mathcal{M}_0 = \{(x, u) : \ x \in \mathcal{A}_0, \ u \in \mathcal{U}_0(x)\} \qquad (8.6)$$

is a closed subset of $E \times F$,

$$\mathcal{M}_* = \{0, 1, \ldots\} \times \mathcal{M}_0, \qquad (8.7)$$

$G_0 : \mathcal{M}_0 \to E$ and that

$$G_*(t, x, u) = G_0(x, u), \ (t, x, u) \in \mathcal{M}. \tag{8.8}$$

Set

$$\mathrm{graph}(G_0) = \{(x, u, y) : \ (x, u) \in \mathcal{M}_0, \ y \in G_0(x, u)\} \tag{8.9}$$

and assume that this set is closed in $E \times F \times E$.

Assume that f_0 is lower semicontinuous, there exist $\bar{x} \in \mathcal{A}_0$, $\bar{u} \in \mathcal{U}_0(\bar{x})$ satisfying

$$\bar{x} = G_0(\bar{x}, \bar{u}), \tag{8.10}$$

$\bar{c} > 0$, f_0 is continuous at $(\bar{x}, \bar{u})$ and such that the following assumptions hold:

(A1) For each $\epsilon > 0$ there exists $\delta > 0$ such that

$$B_E(\bar{x}, \delta) \subset \mathcal{A}_0$$

and that for each $\xi_1, \xi_2 \in B_E(\bar{x}, \delta)$ there exists

$$u \in \mathcal{U}_0(\xi_1) \cap B(\bar{u}, \epsilon)$$

such that

$$\xi_2 = G_0(\xi_1, u).$$

(A2) For any natural number T and any trajectory-control pair

$$(\{x_t\}_{t=0}^T, \{u_t\}_{t=0}^{T-1}) \in X(0, T, \mathcal{M}_*, G_*)$$

we have

$$\sum_{t=0}^{T-1} f_0(x_t, u_t) \geq T f_0(\bar{x}, \bar{u}) - \bar{c}.$$

Assumption (A2) implies that for each trajectory-control pair

$$(\{x_t\}_{t=0}^\infty, \{u_t\}_{t=0}^\infty) \in X(0, \infty, \mathcal{M}_*, G_*)$$

either the sequence

$$\{\sum_{t=0}^{T-1} f_0(x_t, u_t) - T f_0(\bar{x}, \bar{u})\}_{T=1}^\infty$$

is boundedor

$$\lim_{T \to \infty} \left[\sum_{t=0}^{T-1} f_0(x_t, u_t) - T f_0(\bar{x}, \bar{u}) \right] = \infty.$$

A trajectory-control pair

$$(\{x_t\}_{t=0}^{\infty}, \{u_t\}_{t=0}^{\infty}) \in X(0, \infty, \mathcal{M}_*, G_*)$$

is $(\mathcal{M}_0, G_0)$-good if the sequence

$$\{ \sum_{t=0}^{T-1} f_0(x_t, u_t) - T f_0(\bar{x}, \bar{u}) \}_{T=1}^{\infty}$$

is bounded.

Proposition 8.1 *The following properties are equivalent.*

(i) Let $\mathcal{B}$ be the set of all trajectory-control pairs

$$(\{x_t\}_{t=0}^{\infty}, \{u_t\}_{t=0}^{\infty}) \in X(0, \infty, \mathcal{M}_*, G_*)$$

such that $u_t = \bar{u}$, $t = 0, 1, \ldots$. Then for each $(\{x_t\}_{t=0}^{\infty}, \{u_t\}_{t=0}^{\infty}) \in \mathcal{B}$,

$$\lim_{t \to \infty} x_t = \bar{x} \ \text{uniformly on } \mathcal{B}.$$

(ii) If $(\{x_t\}_{t=-\infty}^{\infty} \subset E$ satisfies for each integer t, $(x_t, \bar{u}) \in \mathcal{M}_0$ and $x_{t+1} \in G(x_t, \bar{u}_t)$, then for each integer t, $x_t = \bar{x}$.

Proof Clearly, (i) implies (ii). Assume that (ii) holds. We show that (i) hold. Assume the contrary. Then there exist $\epsilon > 0$ and for each integer $k \geq 1$ there exist a trajectory-control pair $(\{x_t^{(k)}\}_{t=0}^{\infty}, \{u_t^{(k)}\}_{t=0}^{\infty}) \in \mathcal{B}$ and a natural number $n_k \geq k$ such that

$$\rho_E(x_{n_k}^{(k)}, \bar{x}) > \epsilon. \tag{8.11}$$

For each integer $k \geq 1$ and each integer $t \geq -n_k$ define

$$y_t^{(k)} = x_{t+n_k}^{(k)}, \quad v_t^{(k)} = u_{t+n_k}^{(k)} = \bar{u}. \tag{8.12}$$

Extracting a subsequence and re-indexing we may assume without loss of generality that for each integer t there exists

$$y_t = \lim_{k \to \infty} y_t^{(k)}. \tag{8.13}$$

In view of (8.12) and (8.13), for each integer t,

$$(y_t, \bar{u}) \in \mathcal{M}_0, \ \ y_{t+1} \in G_0(y_t, \bar{u}),$$

$$\rho(y_0, \bar{x}) = \lim_{k \to \infty} \rho(y_0^{(k)}, \bar{x}) = \lim_{k \to \infty} \rho(x_{n_k}^{(k)}, \bar{x}) \geq \epsilon.$$

This contradicts (ii). The contradiction we have reached proves that (i) holds. Proposition 8.1 holds.

Proposition 8.2 *Assume that property (i) from Proposition 8.1 holds and that for each $(\mathcal{M}_0, G_0)$-good $(\{x_t\}_{t=0}^{\infty}, \{u_t\}_{t=0}^{\infty}) \in X(0, \infty, \mathcal{M}_*, G_*)$,*

$$\lim_{t \to \infty} u_t = \bar{u}.$$

Then for each $(\mathcal{M}_0, G_0)$-good $(\{x_t\}_{t=0}^{\infty}, \{u_t\}_{t=0}^{\infty}) \in X(0, \infty, \mathcal{M}_*, G_*)$,

$$\lim_{t \to \infty} x_t = \bar{x}.$$

Proof Assume the contrary. Then there exist $\epsilon > 0$ and an $(\mathcal{M}_0, G_0)$-good pair $(\{x_t\}_{t=0}^{\infty}, \{u_t\}_{t=0}^{\infty}) \in X(0, \infty, \mathcal{M}_*, G_*)$ such that

$$\limsup_{t \to \infty} \rho_E(x_t, \bar{x}) > \epsilon. \tag{8.14}$$

In view of our assumptions,

$$\lim_{t \to \infty} \rho_E(u_t, \bar{u}) = 0. \tag{8.15}$$

By (8.14), there exist a strictly increasing sequence of integers $t_k > k$, $k = 1, 2, \ldots$ such that

$$\rho_E(x_{t_k}, \bar{x}) > \epsilon. \tag{8.16}$$

For each integer $k \geq 1$ and each integer $t \geq -t_k$ define,

$$y_t^{(k)} = x_{t+t_k}, \ \ v_t^{(k)} = u_{t+t_k}. \tag{8.17}$$

Extracting a subsequence, using the diagonalization process and re-indexing we may assume without loss of generality that for each integer t there exists

$$y_t = \lim_{k \to \infty} y_t^{(k)}, \ \ \lim_{k \to \infty} v_t^{(k)} = \bar{u}. \tag{8.18}$$

In view of (8.16)–(8.18), for each integer t,

$$(y_t, \bar{u}) \in \mathcal{M}_0, \;\; y_{t+1} \in G_0(y_t, \bar{u}),$$

$$\rho_E(y_0, \bar{x}) = \lim_{k \to \infty} \rho_E(y_0^{(k)}, \bar{x}) = \lim_{k \to \infty} \rho_E(x_{n_k+t}, \bar{x}) \geq \epsilon. \tag{8.19}$$

Property (i), Proposition 8.1 and the relation above imply that for every integer t, $y_t = \bar{x}$. This contradicts (8.19). The contradiction we have reached proves Proposition 8.2.

In this chapter we suppose that the following assumption holds.

(A3) (the asymptotic turnpike property) For each $(\mathcal{M}_0, G_0)$-good trajectory-control pair

$$(\{x_t\}_{t=0}^{\infty}, \{u_t\}_{t=0}^{\infty}) \in X(0, \infty, \mathcal{M}_*, G_*)$$

we have

$$\lim_{t \to \infty} x_t = \bar{x}, \;\; \lim_{t \to \infty} u_t = \bar{u}.$$

Note that (A3) holds for many important infinite horizon optimal control problems. In particular, (A3) holds for the MDP which has the unique minimizing Markov actions.

In this chapter we show that turnpike properties hold and are stable under small perturbations of the objective function and the control maps. In order to meet this goal we introduce the following definitions.

Fix

$$\bar{\lambda} \in (0, 1) \text{ such that } B(\bar{x}, \bar{\lambda}) \subset \mathcal{A}_0. \tag{8.20}$$

For each $\lambda > 0$ denote by $\mathcal{E}(\lambda)$ the collection of all triplets $(\mathcal{A}, \mathcal{U}, G)$, where $\mathcal{A} \subset \{0, 1, \dots\} \times E, \mathcal{U} : \mathcal{A} \to 2^F \setminus \{\emptyset\}$,

$$\mathcal{M} = \text{graph}(\mathcal{U}),$$

$$G : \{(t, x, u) : (t, x) \in \mathcal{A}, \; u \in \mathcal{U}(t, x)\} \to E$$

such the following assumptions hold:

(B1) for each $(t, x) \in \mathcal{A}$, each $u \in \mathcal{U}(t, x)$ and each $y \in G(t, x, u)$,

$$\rho_{E \times F \times E}((x, u, y), \text{graph}(G_0)) \leq \lambda;$$

(B2) for each integer $t \geq 0$,

$$\{(x, u, y) : (x, u) \in \mathcal{M}_0, \; y \in G_0(x, y),$$

$$\rho_E(x, \bar{x}), \ \rho_E(y, \bar{x}) \leq \bar{\lambda}, \ \rho_F(u, \bar{u}) \leq \bar{\lambda}\}$$

$$\subset \{(x, u, y) : \ (t, x) \in \mathcal{A}, \ u \in \mathcal{U}(t, x), \ y \in G(t, x, u)\}, \ t = 0, 1, \ldots.$$

By (A1), there exists

$$\widehat{\lambda} \in (0, \bar{\lambda}) \tag{8.21}$$

such that the following property holds:

(B3) for each $\xi_1, \xi_2 \in B_E(\bar{x}, \widehat{\lambda})$ there exists

$$u \in \mathcal{U}_0(\xi_1) \cap B_F(\bar{u}, \widehat{\lambda})$$

such that

$$\xi_2 \in G_0(\xi_1, u).$$

Assume that $\mathcal{A}$ is a nonempty subset of $\{0, 1, \ldots, \} \times E, \mathcal{U} : \mathcal{A} \to 2^F$ is a point to set mapping,

$$\mathcal{M} = \mathrm{graph}(\mathcal{U}) = \{(t, x, u) : \ (t, x) \in \mathcal{A}, \ u \in \mathcal{U}(t, x)\},$$

$G : \mathcal{M} \to E$. Let $0 \leq T_1 < T_2$ be integers.

Denote by $Y(T_1, T_2, \mathcal{M}, G)$ the set of all $x \in X$ for which there exists $(\{x_t\}_{t=T_1}^{T_2}, \{u_t\}_{t=T_1}^{T_2-1}) \in X(T_1, T_2, \mathcal{M}, G)$ such that $x_{T_1} = \bar{x}$ and $x_{T_2} = x$.

Denote by $\bar{Y}(T_1, T_2, \mathcal{M}, G)$ the set of all $x \in X$ for which there exists $(\{x_t\}_{t=T_1}^{T_2}, \{u_t\}_{t=T_1}^{T_2-1}) \in X(T_1, T_2, \mathcal{M}, G)$ such that $x_{T_1} = x$ and $x_{T_2} = \bar{x}$.

In this chapter we prove the following stability results. The first of them shows that the strong turnpike phenomenon is stable under small perturbations of the objective function and the control maps.

Theorem 8.3 *Let $\epsilon \in (0, \widehat{\lambda})$ and let l_1, l_2 be natural numbers. Then there exist $\delta \in (0, \epsilon)$ and a natural number $L > l_1 + l_2$ such that for each integer $T > 2L$, each $f \in \mathfrak{M}$ satisfying*

$$\|f - f_*\| \leq \delta,$$

each

$$(\mathcal{A}, \mathcal{U}, G) \in \mathcal{E}(\delta), \ \mathcal{M} = graph(\mathcal{U}),$$

each $(\{x_t\}_{t=0}^{T}, \{u_t\}_{t=0}^{T-1}) \in X(0, T, \mathcal{M}, G)$ satisfying

$$x_0 \in \bar{Y}(0, l_1, \mathcal{M}, G)$$

and such that at least one of the following conditions holds:

(a)

$$x_T \in Y(T - l_2, T, \mathcal{M}, G),$$

$$\sum_{t=0}^{T-1} f(t, x_t, u_t) \le U_{\mathcal{M},G}^{f}(0, T, x_0, x_T) + \delta;$$

(b)

$$\sum_{t=0}^{T-1} f(t, x_t, u_t) \le U_{\mathcal{M},G}^{f}(0, T, x_0) + \delta$$

there exist integers $\tau_1 \in [0, L]$, $\tau_2 \in [T - L, T]$ such that

$$\rho_E(x_t, \bar{x}) \le \epsilon \ \textit{for all } t \in \{\tau_1, \ldots, \tau_2\},$$

$$\rho_F(u_t, \bar{u}) \le \epsilon \ \textit{for all } t \in \{\tau_1, \ldots, \tau_2\} \setminus \{\tau_2\},$$

Moreover if $\rho(x_0, \bar{x}) \le \delta$, then $\tau_1 = 0$ and if $\rho(x_T, \bar{x}) \le \delta$, then $\tau_2 = T$.

Our second result shows that the weak turnpike phenomenon is stable under small perturbations of the objective function and the control maps.

Theorem 8.4 *Let ϵ, M be positive numbers and let l_1, l_2 be natural numbers. Then there exist $\delta \in (0, \widehat{\lambda})$ and a natural number $L > l_1 + l_2$ such that for each integer $T > L$, each $f \in \mathfrak{M}$ satisfying*

$$\|f - f_*\| \le \delta,$$

each

$$(\mathcal{A}, \mathcal{U}, G) \in \mathcal{E}(\delta), \ \mathcal{M} = graph(\mathcal{U}),$$

each $(\{x_t\}_{t=0}^{T}, \{u_t\}_{t=0}^{T-1}) \in X(0, T, \mathcal{M}, G)$ satisfying

$$x_0 \in \bar{Y}(0, l_1, \mathcal{M}, G)$$

and such that at least one of the following conditions holds:

(a)

$$x_T \in Y(T - l_2, T, \mathcal{M}, G),$$

$$\sum_{t=0}^{T-1} f(t, x_t, u_t) \le U_{\mathcal{M},G}^{f}(0, T, x_0, x_T) + M;$$

(b)

$$\sum_{t=0}^{T-1} f(t, x_t, u_t) \le U_{\mathcal{M},G}^{f}(0, T, x_0) + M$$

the inequality

$$Card(\{t \in \{0, \dots, T\} : \ \max\{\rho_E(x_t, \bar{x}), \ \rho_F(u_t, \bar{u}) > \epsilon\}) \le L$$

holds.

In Theorem 8.3 the strong turnpike property is established for δ-approximate solutions of our optimal control problem where δ is sufficiently small while in Theorem 8.4 the weak turnpike property is established for M-approximate solutions where M is a given positive constant which is not necessarily small.

8.2   Three Lemmata

In order to prove our stability results we need the following useful lemmas.

Lemma 8.5 *Let $\epsilon > 0$ and $M_0 > 0$. Then there exists a natural number $T \ge 2$ such that for each $(\{x_t\}_{t=0}^{T}, \{u_t\}_{t=0}^{T-1}) \in X(0, T, \mathcal{M}_*, G_*)$ which satisfies*

$$\sum_{t=0}^{T-1} f_0(x_t, u_t) \le T f_0(\bar{x}, \bar{u}) + M_0$$

the relation

$$\min\{\rho_E(x_t, \bar{x}) + \rho_F(u_t, \bar{u}) : \ t = 1, \dots, T-1\} \le \epsilon$$

holds.

Proof Let us assume that the lemma does not hold. Then for each natural number $k \ge 2$ there exists

$$(\{x_t^{(k)}\}_{t=0}^{k}, \{u_t^{(k)}\}_{t=0}^{k-1}) \in X(0, k, \mathcal{M}_*, G_*)$$

which satisfies

$$\sum_{t=0}^{k-1} f_0(x_t^{(k)}, u_t^{(k)}) \le k f_0(\bar{x}, \bar{u}) + M_0, \tag{8.22}$$

$$\rho_E(x_t^{(k)}, \bar{x}) + \rho_F(u_t^{(k)}, \bar{u}) > \epsilon \text{ for all integers } t = 1, \dots, k-1. \tag{8.23}$$

Let $k \ge 2$ be an integer. By (8.22) and (A2) for each integer j satisfying $0 < j < k$

$$\sum_{t=0}^{j-1} f_0(x_t^{(k)}, u_t^{(k)}) = \sum_{t=0}^{k-1} f_0(x_t^{(k)}, u_t^{(k)}) - \sum_{t=j}^{k-1} f_0(x_t^{(k)}, u_t^{(k)})$$

$$\le k f_0(\bar{x}, \bar{u}) + M_0 - \sum_{t=j}^{k-1} f_0(x_t^{(k)}, u_t^{(k)})$$

$$\le k f_0(\bar{x}, \bar{u}) + M_0 - (k-j) f_0(\bar{x}, \bar{u}) + \bar{c}.$$

This inequality implies that for each integer $k \ge 2$ and each $j \in \{1, \dots, k\}$

$$\sum_{t=0}^{j-1} f_0(x_t^{(k)}, u_t^{(k)}) \le j f_0(\bar{x}, \bar{u}) + \bar{c} + M_0. \tag{8.24}$$

There exists a strictly increasing sequence of natural numbers $\{k_i\}_{i=1}^{\infty}$ such that for each integer $t \ge 0$ there exists

$$x_t = \lim_{i \to \infty} x_t^{(k_i)}, \quad u_t = \lim_{i \to \infty} u_t^{(k_i)}. \tag{8.25}$$

Since the graph of G_0 is closed $(\{x_t\}_{t=0}^{\infty}, \{u_t\}_{t=0}^{\infty}) \in X(0, \infty, \mathcal{M}_*, G_*)$. By (8.23) and (8.25), for all integers $t \ge 1$,

$$\rho_E(x_t, \bar{x}) + \rho_F(u_t, \bar{u}) \ge \epsilon.$$

It follows from lower semicontinuity of f_0 and (8.24) that for each integer $T \ge 1$ we have

$$\sum_{t=0}^{T-1} f_0(x_t, u_t) \le T f_0(\bar{x}, \bar{u}) + M_0 + \bar{c}.$$

This implies that $(\{x_t\}_{t=0}^{\infty}, \{u_t\}_{t=0}^{\infty})$ is $(\mathcal{M}_0, G_0)$-good. In view of (A3),

$$\lim_{t \to \infty} \rho_E(x_t, \bar{x}) = \lim_{t \to \infty} \rho_F(u_t, \bar{u}) = 0.$$

The contradiction we have reached proves Lemma 8.5.

Lemma 8.6 *Let $\epsilon > 0$. Then there exists $\delta > 0$ such that for each integer $T \geq 1$ and each $(\{x_t\}_{t=0}^{T}, \{u_t\}_{t=0}^{T-1}) \in X(0, T, \mathcal{M}_*, G_*)$ which satisfies*

$$\rho_E(x_0, \bar{x}), \ \rho_E(x_T, \bar{x}) \leq \delta,$$

$$\sum_{t=0}^{T-1} f_0(x_t, u_t) \leq U_{\mathcal{M}_*, G_*}^{f_*}(0, T, x_0, x_T) + \delta$$

the inequalities $\rho_E(x_t, \bar{x}), \ \rho_F(u_t, \bar{x}) \leq \epsilon$ holds for all $t = 0, \ldots, T - 1$.

Proof Since f_0 is continuous at $(\bar{x}, \bar{u})$ for each natural number k there exists

$$\gamma_k \in (0, 2^{-k}\widehat{\lambda}] \tag{8.26}$$

such that

$$|f_0(x, u) - f_0(\bar{x}, \bar{u})| \leq 2^{-k} \tag{8.27}$$

for each $x \in B_E(\bar{x}, \gamma_k)$ and each $u \in B_F(\bar{u}, \gamma_k)$.

By (A1), for each integer $k \geq 1$, there exists

$$\delta_k \in (0, \gamma_k) \tag{8.28}$$

such that the following property holds:

(i) for each $\xi_1, \xi_2 \in B(\bar{x}, \delta_k)$ there exist

$$u \in \mathcal{U}_0(\xi_1) \cap B_F(\bar{u}, \gamma_k)$$

such that

$$\xi_2 \in G_0(\xi_1, u).$$

Assume that the lemma is wrong. Then for each natural number k there exist an integer $T_k \geq 1$ and

$$(\{x_t^{(k)}\}_{t=0}^{T_k}, \{u_t^{(k)}\}_{t=0}^{T_k-1}) \in X(0, T_k, \mathcal{M}_*, G_*)$$

such that

$$\rho_E(x_0^{(k)}, \bar{x}), \ \rho_E(x_{T_k}^{(k)}, \bar{x}) \leq \delta_k, \tag{8.29}$$

$$\sum_{t=0}^{T-1} f_0(x_t^{(k)}, u_t^{(k)}) \leq U_{\mathcal{M}_*, G_*}^{f_*}(0, T_k, x_0^{(k)}, x_T^{(k)}) + \delta_k \tag{8.30}$$

$$\max\{\rho_E(x_t^{(k)}, \bar{x}) + \rho_F(u_t^{(k)}, \bar{u}) : \ t = 0, \ldots, T_k - 1\} > \epsilon. \tag{8.31}$$

Let $k \geq 1$ be an integer. There are two cases: $T_k \geq 2$; $T_k = 1$. Assume that $T_k \geq 2$. Set

$$z_0 = x_0^{(k)}. \tag{8.32}$$

Property (i) and (8.29) and (8.32) imply that there exist

$$v_0 \in \mathcal{U}_0(z_0) \cap B_F(\bar{u}, \gamma_k) \tag{8.33}$$

such that

$$\bar{x} \in G_0(z_0, v_0). \tag{8.34}$$

For each $t \in \{1, \ldots, T_k - 1\}$ set

$$z_t = \bar{x} \tag{8.35}$$

and for each $t \in \{1, \ldots, T_k - 1\} \setminus \{T_k - 1\}$ set

$$v_t = \bar{u}. \tag{8.36}$$

Property (i) and (8.29), (8.35) imply that there exists

$$v_{T_k-1} \in \mathcal{U}_0(z_{T_k-1}) \cap B_F(\bar{u}, \gamma_k) = \mathcal{U}_0(\bar{x}) \cap B_F(\bar{u}, \gamma_k) \tag{8.37}$$

such that

$$x_{T_k}^{(k)} \in G_0(\bar{x}, v_{T_k-1}). \tag{8.38}$$

Set

$$z_{T_k} = x_{T_k}^{(k)}. \tag{8.39}$$

By (8.10), (8.32) and (8.39),

$$(\{z_t^{(k)}\}_{t=0}^{T_k}, \{v_t^{(k)}\}_{t=0}^{T_k-1}) \in X(0, T_k, \mathcal{M}_*, G_*).$$

By (8.30), (8.32), (8.35), (8.36) and (8.39),

$$\sum_{t=0}^{T_k-1} f_0(x_t^{(k)}, u_t^{(k)}) \le \sum_{t=0}^{T_k-1} f_0(z_t, u_t) + \delta_k$$

$$\le T_k f_0(\bar{x}, \bar{u}) + \delta_k + |f_0(\bar{x}, \bar{u}) - f_0(x_0^{(k)}, v_0)| + |f_0(\bar{x}, \bar{u}) - f_0(\bar{x}, v_{T_k-1})|. \tag{8.40}$$

It follows from (8.27), (8.29), (8.33) and (8.37) that

$$|f_0(\bar{x}, \bar{u}) - f_0(x_0^{(k)}, v_0)| \le 2^{-k}, \tag{8.41}$$

$$|f_0(\bar{x}, \bar{u}) - f_0(\bar{x}, v_{T_k-1})| \le 2^{-k}. \tag{8.42}$$

Equations (8.26), (8.28), (8.41) and (8.42) imply that

$$\sum_{t=0}^{T_k-1} f_0(x_t^{(k)}, u_t^{(k)}) \le T_k f_0(\bar{x}, \bar{u}) + 3 \cdot 2^{-k}. \tag{8.43}$$

Assume that $T_k = 1$. Property (i) and (8.29) imply that there exist

$$v_0 \in \mathcal{U}_0(x_0^{(k)}) \cap B_F(\bar{u}, \gamma_k) \tag{8.44}$$

such that

$$x_1^{(k)} \in G_0(x_0^{(k)}, v_0). \tag{8.45}$$

In view of (8.44) and (8.45),

$$(\{x_0^{(k)}, x_1^{(k)}\}, \{v_0\}) \in X(0, 1, \mathcal{M}_*, G_*).$$

By (8.26)–(8.30), (8.44) and (8.45),

$$f_0(x_0^{(k)}, u_0^{(k)}) \le f_0(x_k^{(0)}, v_0) + \delta_k \le f_0(\bar{x}, \bar{u}) + 2^{-k+1}.$$

Thus in both cases (8.43) holds. Define

$$\tilde{x}_0^{(k)} = \bar{x}. \tag{8.46}$$

Property (i) and (8.29) imply that there exists

$$\tilde{u}_0^{(k)} \in \mathcal{U}_0(\bar{x}) \cap B_F(\bar{u}, \gamma_k) \tag{8.47}$$

such that

$$x_0^{(k)} \in G_0(\bar{x}, \tilde{u}_0^{(k)}). \tag{8.48}$$

Set

$$\tilde{x}_t^{(k)} = x_{t-1}^{(k)}, \ t = 1, \ldots, T_k + 1, \ \tilde{x}_{T_k+2}^{(k)} = \bar{x}, \tag{8.49}$$

$$\tilde{u}_t^{(k)} = u_{t-1}(k), \ t = 1, \ldots, T_k. \tag{8.50}$$

Property (i) and (8.29), (8.49) imply that there exists

$$\tilde{u}_{T_k+1}^{(k)} \in \mathcal{U}_0(\tilde{x}_{T_k+1}^{(k)}) \cap B_F(\bar{u}, \gamma_k) \tag{8.51}$$

such that

$$\bar{x} \in G_0(\tilde{x}_{T_k+1}^{(k)}, \tilde{u}_{T_k+1}^{(k)}). \tag{8.52}$$

By (8.46) and (8.52),

$$(\{\tilde{x}^{(k)}\}_{k=0}^{T_k+2}, \{\tilde{u}^{(k)}\}_{k=0}^{T_k+1}) \in X(0, T_k + 2, \mathcal{M}_*, G_*).$$

By (8.43), (8.49) and (8.50),

$$\sum_{t=0}^{T_k+1} f_0(\tilde{x}_t^{(k)}, \tilde{u}_t^{(k)}) = f_0(\tilde{x}_t^{(0)}, \tilde{u}_0^{(0)})$$

$$+ \sum_{t=0}^{T_k-1} f_0(x_t^{(k)}, u_t^{(k)}) + f_0(\tilde{x}_{T_k+1}^{(k)}, \tilde{u}_{T_k+1}^{(k)})$$

$$\leq T_k f_0(\bar{x}, \bar{u}) + 3 \cdot 2^{-k} + f_0(\tilde{x}_0^{(k)}, \tilde{u}_0^{(k)}) + f_0(\tilde{x}_{T_k+1}^{(k)}, \tilde{u}_{T_k+1}^{(k)}). \tag{8.53}$$

It follows from (8.27), (8.28), (8.46), (8.47), (8.49) and (8.51) that

$$|f_0(\bar{x}, \bar{u}) - f_0(\tilde{x}_t^{(k)}, \tilde{u}_t^{(k)})| \leq 2^{-k}, \ t \in \{0, T_k + 1\}. \tag{8.54}$$

By (8.53) and (8.54),

$$\sum_{t=0}^{T_k+1} f_0(\tilde{x}_t^{(k)}, \tilde{u}_t^{(k)}) \leq (T_k + 2) f_0(\bar{x}, \bar{u}) + 5 \cdot 2^{-k}. \tag{8.55}$$

In view of (8.31), (8.49) and (8.50),

$$\max\{\rho_E(\tilde{x}_t^{(k)}, \bar{x}) + \rho_F(\tilde{u}_t^{(k)}, \bar{u}) : \ t = 1, \ldots, T_k\} > \epsilon. \tag{8.56}$$

Set

$$S_0 = 0, \ \ S_1 = T_1 + 1,$$

$$S_k = \sum_{i=1}^{k} (T_i + 2) - 1 \text{ for all integers } k \geq 1. \tag{8.57}$$

Define

$$(\{x_t\}_{t=0}^{\infty}, \{u_t\}_{t=0}^{\infty}) \in X(0, \infty, \mathcal{M}_*, G_*)$$

as follows:

$$x_t = \tilde{x}_t^{(1)}, \ u_t = \tilde{u}_t^{(1)}, \ t = 0, \ldots, T_1 + 1, \tag{8.58}$$

$$x_t = \tilde{x}_{i-1}^{(k+1)}, \ u_t = \tilde{u}_{i-1}^{(k+1)} \tag{8.59}$$

for each integer $k \geq 1$, each $i \in \{1, \ldots, T_{k+1} + 2\}$ and $t = S_k + i$.

Equations (8.56) and (8.59) imply that

$$\limsup_{t \to \infty} (\rho(x_t, \bar{x}) + \rho(u_t, \bar{u})) \geq \epsilon. \tag{8.60}$$

By (8.55) and (8.57)–(8.59), for any integer $p \geq 2$,

$$\sum_{t=0}^{S_p} f_0(x_t, u_t) = \sum_{k=1}^{p} \left(\sum_{t=0}^{T_k+1} f_0(\tilde{x}_t, \tilde{u}_t) \right)$$

$$\leq \sum_{k=1}^{p} ((T_k + 2) f_0(\bar{x}, \bar{u}) + 5 \cdot 2^{-k}) \leq (S_p + 1) f_0(\bar{x}, \bar{u}) + 5.$$

Thus $(\{x_t\}_{t=0}^{\infty}, \{u_t\}_{t=0}^{\infty})$ is $(\mathcal{M}_*, G_*)$-good and in view of (A3),

$$\lim_{t \to \infty} \rho(x_t, \bar{x}) = 0 \ \lim_{t \to \infty} \rho(u_t, \bar{u}) = 0.$$

This contradicts (8.60). The contradiction we have reached proves Lemma 8.6.

Lemma 8.7 *Let $\epsilon > 0$ and $M_0 > 0$. Then there exists a natural number $T_0 \geq 2$ such that for each integer $T \geq T_0$, each*

$$(\{x_t\}_{t=0}^T, \{u_t\}_{t=0}^{T-1}) \in X(0, T, \mathcal{M}_*, G_*)$$

which satisfies

$$\sum_{t=0}^{T-1} f_0(x_t, u_t) \leq T f_0(\bar{x}, \bar{u}) + M_0 \tag{8.61}$$

and each integer $s \in [0, T - T_0]$ the inequality

$$\min\{\rho_E(x_i, \bar{x}) + \rho_F(u_i, \bar{u}) : i = s + 1, \ldots, s + T_0 - 1\} \leq \epsilon \tag{8.62}$$

holds.

Proof By Lemma 8.5 there is a natural number $T_0 \geq 2$ such that the following property holds:

(i) for each $(\{x_t\}_{t=0}^T, \{u_t\}_{t=0}^{T-1}) \in X(0, T, \mathcal{M}_*, G_*)$ which satisfies

$$\sum_{t=0}^{T-1} f_0(x_t, u_t) \leq T f_0(\bar{x}, \bar{u}) + M_0 + 2\bar{c}$$

the relation

$$\min\{\rho_E(x_t, \bar{x}) + \rho_F(u_t, \bar{u}) : \ t = 1, \ldots, T_0 - 1\} \leq \epsilon$$

holds.

Let an integer $T \geq T_0$, $(\{x_t\}_{t=0}^T, \{u_t\}_{t=0}^{T-1}) \in X(0, T, \mathcal{M}_*, G_*)$, (8.61) holds, and an integer $s \in [0, T - T_0]$. It follows from (8.61) and (A2) that

$$\sum_{t=s}^{s+T_0-1} f_0(x_t, u_t) \leq T_0 f_0(\bar{x}, \bar{u}) \leq M_0 + 2\bar{c}.$$

By the inequality above and (i), (8.62) holds. Lemma 8.7 is proved.

8.3   Auxiliary Results for Perturbed Problems

Lemma 8.8 *Let ϵ be a positive number and let L be a natural number. Then there exists $\delta > 0$ such that for each*

$$(\mathcal{A}, \mathcal{U}, G) \in \mathcal{E}(\delta), \ \mathcal{M} = \text{graph}(\mathcal{U}),$$

each $f \in \mathfrak{M}$ satisfying

$$\|f - f_*\| \leq \delta,$$

each $(\{x_t\}_{t=0}^{L}, \{u_t\}_{t=0}^{L-1}) \in X(0, L, \mathcal{M}, G)$ *there exists* $(\{\tilde{x}_t\}_{t=0}^{L}, \{\tilde{u}_t\}_{t=0}^{L-1}) \in$
$X(0, L, \mathcal{M}_*, G_*)$ *such that*

$$\rho_E(x_t, \tilde{x}_t) \leq \epsilon, \ t = 0, \ldots, L, \ \rho_F(u_t, \tilde{u}_t) \leq \epsilon, \ t = 0, \ldots, L-1$$

and

$$\sum_{t=0}^{L-1} f_0(\tilde{x}_t, \tilde{u}_t) \leq \sum_{t=0}^{L-1} f(t, x_t, u_t) + \epsilon.$$

Proof Assume that the lemma does not hold. Then for each natural number k there
exist

$$(\mathcal{A}_k, \mathcal{U}_k, G_k) \in \mathcal{E}(k^{-1}\bar{\lambda}), \ \mathcal{M}_k = \text{graph}(\mathcal{U}_k), \tag{8.63}$$

$f_k \in \mathfrak{M}$ satisfying

$$\|f_k - f_*\| \leq k^{-1}\bar{\lambda}, \tag{8.64}$$

$$(\{x_t^{(k)}\}_{t=0}^{L}, \{u_t^{(k)}\}_{t=0}^{L-1}) \in X(0, L, \mathcal{M}_k, G_k)$$

such that the following property holds:

(P1)   if

$$(\{x_t\}_{t=0}^{L}, \{u_t\}_{t=0}^{L-1}) \in X(0, L, \mathcal{M}_*, G_*)$$

satisfies

$$\rho_E(x_t, x_t^{(k)}) \leq \epsilon, \ t = 0, \ldots, L, \ \rho_F(u_t, u_t^{(k)}) \leq \epsilon, \ t = 0, \ldots, L-1,$$

then

$$\sum_{t=0}^{L-1} f_0(x_t, u_t) > \sum_{t=0}^{L-1} f_k(t, x_t^{(k)}, u_t^{(k)}) + \epsilon.$$

Extracting subsequences and re-indexing, if necessary, we may assume
without loss of generality that there exist

$$x_t = \lim_{k \to \infty} x_t^{(k)}, \ t = 0, \ldots, L, \ u_t = \lim_{k \to \infty} u_t^{(k)}, \ t = 0, \ldots, L-1.$$

$$\tag{8.65}$$

By (B1), (8.63) and (8.65), for each integer $t \in \{0, \ldots, L - 1\}$ and each natural number k,

$$\rho_{E \times F \times E}((x_t, u_t, x_{t+1}), \operatorname{graph}(G_0))$$

$$\leq \rho_{E \times F \times E}((x_t, u_t, x_{t+1}), (x_t^{(k)}, u_t^{(k)}, x_{t+1}^{(k)}))$$

$$+ \rho_{E \times F \times E}((x_t^{(k)}, u_t^{(k)}, x_{t+1}^{(k)}), \operatorname{graph}(G_0))$$

$$\leq \rho_{E \times F \times E}((x_t, u_t, x_{t+1}), (x_t^{(k)}, u_t^{(k)}, x_{t+1}^{(k)})) + k^{-1} \to 0 \text{ as } k \to \infty$$

and

$$(x_t, u_t, x_{t+1}) \in \operatorname{graph}(G_0),$$

$$(\{x_t\}_{t=0}^{L}, \{u_t\}_{t=0}^{L-1}) \in X(0, L, \mathcal{M}_*, G_*).$$

By (8.65) and lower semicontinuity of the function f_0, there exists a natural number k_0 such that for each integer $k \geq k_0$,

$$\rho_E(x_t, x_t^{(k)}) \leq \epsilon, \ t = 0, \ldots, L, \ \rho_F(u_t, u_t^{(k)}) \leq \epsilon, \ t = 0, \ldots, L - 1 \qquad (8.66)$$

and

$$\sum_{t=0}^{L-1} f_0(x_t, u_t) \leq \sum_{t=0}^{L-1} f_0(x_t^{(k)}, u_t^{(k)}) + \epsilon/4. \qquad (8.67)$$

Choose a natural number $k_1 > k_0$ such that

$$16 k_1^{-1} L < \epsilon. \qquad (8.68)$$

Assume that an integer $k \geq k_1$. Then (8.66) and (8.67) hold. In view of (8.64), (8.67) and (8.68),

$$\sum_{t=0}^{L-1} f_0(x_t, u_t) \leq \sum_{t=0}^{L-1} f_0(x_t^{(k)}, u_t^{(k)}) + \epsilon/4$$

$$\leq \sum_{t=0}^{L-1} f_k(t, x_t^{(k)}, u_t^{(k)}) + \epsilon/4 + L \|f_* - f_k\|$$

$$\leq \sum_{t=0}^{L-1} f_k(t, x_t^{(k)}, u_t^{(k)}) + \epsilon/4 + Lk^{-1}$$

$$\leq \sum_{t=0}^{L-1} f_k(t, x_t^{(k)}, u_t^{(k)}) + \epsilon/2.$$

When combined with (8.66) this contradicts property (P1). The contradiction we have reached proves Lemma 8.8.

Lemma 8.9 *Let ϵ be a positive number and let L be a natural number. Then there exists $\delta > 0$ such that for each*

$$(\mathcal{A}, \mathcal{U}, G) \in \mathcal{E}(\delta), \ \mathcal{M} = graph(\mathcal{U}), \tag{8.69}$$

each $f \in \mathfrak{M}$ satisfying

$$\|f - f_*\| \leq \delta, \tag{8.70}$$

each integer $s \geq 0$ and each $(\{x_t\}_{t=s}^{s+L}, \{u_t\}_{t=s}^{s+L-1}) \in X(s, s+L, \mathcal{M}, G)$ there exists $(\{\widehat{x}_t\}_{t=s}^{s+L}, \{\widehat{u}_t\}_{t=s}^{s+L-1}) \in X(s, s+L, \mathcal{M}_*, G_*)$ such that

$$\rho_E(x_t, \widehat{x}_t) \leq \epsilon, \ t = s, \dots, s+L, \ \rho_F(u_t, \widehat{u}_t) \leq \epsilon, \ t = s, \dots, s+L-1$$

and

$$\sum_{t=s}^{s+L-1} f_0(\widehat{x}_t, \widehat{u}_t) \leq \sum_{t=s}^{s+L-1} f(t, x_t, u_t) + \epsilon.$$

Proof Let $\delta > 0$ be as guaranteed by Lemma 8.8, (8.69) hold, $f \in \mathfrak{M}$ satisfy (8.70), $\mathcal{M} = graph(\mathcal{U})$, $s \geq 0$ be an integer, $(\{x_t\}_{t=s}^{s+L}, \{u_t\}_{t=s}^{s+L-1}) \in X(s, s+L, \mathcal{M}, G)$. Define

$$\tilde{x}_t = x_{t+s}, \ t = 0, \dots, L, \ \tilde{u}_t = u_{t+s}, \ t = 0, \dots, L-1,$$

$$\tilde{f}(t, x, u) = f(t+s, x, u), \ (x, u) \in E \times G, \ t = 0, 1, \dots,$$

$$\tilde{\mathcal{A}} = \{(t, x) : \ (t+s, x) \in \mathcal{A}\},$$

$$\tilde{\mathcal{U}}(t, x) = \mathcal{U}(t+s, x), \ (t, x) \in \tilde{\mathcal{A}},$$

$$\tilde{\mathcal{M}} = \ graph(\tilde{\mathcal{U}}),$$

$$\tilde{G}(t, x, u) = G(t+s, x, u), \ (t, x) \in \tilde{\mathcal{A}}, \ u \in \tilde{\mathcal{U}}(t, x).$$

(B1), (B2), (8.69) and (8.70) imply that

$$\|\tilde{f} - f_*\| \le \delta,$$

$$(\tilde{A}, \tilde{U}, \tilde{G}) \in \mathcal{E}(\delta).$$

Lemma 8.8 , the relation above and the choice of δ imply that that there exists

$$(\{\bar{x}_t\}_{t=0}^{L}, \{\bar{u}_t\}_{t=0}^{L-1}) \in X(0, L, \mathcal{M}_*, G_*)$$

such that

$$\rho_E(\bar{x}_t, x_{t+s}) = \rho_E(\tilde{x}_t, \bar{x}_t) \le \epsilon, \ t = 0, \ldots, L,$$

$$\rho_F(\tilde{u}_t, \bar{u}_t) = \rho_F(\bar{u}_t, u_{t+s}) \le \epsilon, \ t = 0, \ldots, L - 1,$$

$$\sum_{t=0}^{L-1} f_0(\bar{x}_t, \bar{u}_t) \le \sum_{t=0}^{L-1} f(t, \tilde{x}_t, \tilde{u}_t) + \epsilon$$

$$= \sum_{t=s}^{s+L-1} f(t, x_t, u_t) + \epsilon.$$

Set

$$\widehat{x}_t = \bar{x}_{t-s}, \ t = s, \ldots, s + L, \ \widehat{u}_t = \bar{u}_{t-s}, \ t = s, \ldots, s + L - 1.$$

We have $(\{\widehat{x}_t\}_{t=s}^{s+L}, \{\widehat{u}_t\}_{t=s}^{s+L-1}) \in X(s, s + L, \mathcal{M}_*, G_*)$,

$$\rho_E(x_t, \widehat{x}_t) \le \epsilon, \ t = s, \ldots, s + L, \ \rho_F(u_t, \widehat{u}_t) \le \epsilon, \ t = s, \ldots, s + L - 1$$

and

$$\sum_{t=s}^{s+L-1} f_0(\widehat{x}_t, \widehat{u}_t) \le \sum_{t=s}^{s+L-1} f(t, x_t, u_t) + \epsilon.$$

Lemma 8.9 is proved.

Lemma 8.10 *Let $\epsilon \in (0, 1)$, $M > 0$. Then there exists a natural number $L \ge 2$ such that for each integer $\tilde{L} > L + 1$ there exists $\delta > 0$ such that the following assertion holds.*
Assume that $T \in \{L + 1, \ldots, \tilde{L}\}$,

$$(A, U, G) \in \mathcal{E}(\delta), \ \mathcal{M} = graph(U), \tag{8.71}$$

$f \in \mathfrak{M}$,

$$\|f - f_*\| \leq \delta, \tag{8.72}$$

$s \geq 0$ *is an integer,*

$$(\{x_t\}_{t=s}^{s+T}, \{u_t\}_{t=s}^{s+T-1}) \in X(s, s+T, \mathcal{M}, G), \tag{8.73}$$

$$\sum_{t=s}^{s+T-1} f(t, x_t, u_t) \leq \sum_{t=s}^{s+T-1} f(t, \bar{x}, \bar{u}) + M_0. \tag{8.74}$$

Then

$$\min\{\rho_E(x_t, \bar{x}) + \rho_F(u_t, \bar{u}) : t = s+1, \ldots, s+T-1\} \leq \epsilon.$$

Proof By Lemma 8.7, there exists a natural number $L \geq 2$ such that the following property holds:

(P2) for each integer $T \geq L$, each

$$(\{x_t\}_{t=0}^{T}, \{u_t\}_{t=0}^{T-1}) \in X(0, T, \mathcal{M}_*, G_*)$$

which satisfies

$$\sum_{t=0}^{T-1} f_0(x_t, u_t) \leq T f_0(\bar{x}, \bar{u}) + M + 4$$

and each integer $s \in [0, T - L]$ we have

$$\min\{\rho_E(x_i, \bar{x}) + \rho_F(u_i, \bar{u}) : i = s+1, \ldots, s+L-1\} \leq \epsilon/4.$$

Let $\tilde{L} > L + 1$ be an integer. Lemma 8.9 implies that there exists $\delta_0 > 0$ such that the following property holds:

(P3) for each integer $T \in \{L, \ldots, \tilde{L}\}$, each

$$(\mathcal{A}, \mathcal{U}, G) \in \mathcal{E}(\delta_0), \quad \mathcal{M} = \mathrm{graph}(\mathcal{U}),$$

each $f \in \mathfrak{M}$ satisfying

$$\|f - f_*\| \leq \delta_0,$$

each integer $s \geq 0$, each $(\{x_t\}_{t=s}^{s+T}, \{u_t\}_{t=s}^{s+T-1}) \in X(s, s+T, \mathcal{M}, G)$ there exists $(\{\widehat{x}_t\}_{t=s}^{s+T}, \{\widehat{u}_t\}_{t=s}^{s+T-1}) \in X(s, s+T, \mathcal{M}_*, G_*)$ such that

$$\rho_E(x_t, \widehat{x}_t) \le \epsilon/4, \ t = s, \ldots, s+T, \ \ \rho_F(u_t, \widehat{u}_t) \le \epsilon/4, \ t = s, \ldots, s+T-1$$

and

$$\sum_{t=s}^{s+T-1} f_0(\widehat{x}_t, \widehat{u}_t) \le \sum_{t=s}^{s+T-1} f(t, x_t, u_t) + \epsilon/4.$$

Set

$$\delta = \min\{\delta_0, \ \epsilon \tilde{L}^{-1}\}. \tag{8.75}$$

Assume that an integer $T \in [L + 1, \tilde{L}]$, (8.71) holds,

$$\mathcal{M} = \mathrm{graph}(\mathcal{U}), \ f \in \mathfrak{M},$$

(8.72) is true, $s \ge 0$ is an integer, (8.73), (8.74) hold. Property (P3) and (8.71)–(8.75) imply that there exists $(\{\widehat{x}_t\}_{t=s}^{s+T}, \{\widehat{u}_t\}_{t=s}^{s+T-1}) \in X(s, s + T, \mathcal{M}_*, G_*)$ such that

$$\rho_E(x_t, \widehat{x}_t) \le \epsilon/4, \ t = s, \ldots, s + T, \tag{8.76}$$

$$\rho_F(u_t, \widehat{u}_t) \le \epsilon/4, \ t = s, \ldots, s + T - 1, \tag{8.77}$$

and

$$\sum_{t=s}^{s+T-1} f_0(\widehat{x}_t, \widehat{u}_t) \le \sum_{t=s}^{s+T-1} f(t, x_t, u_t) + \epsilon/4. \tag{8.78}$$

By (8.72), (8.74), (8.75) and (8.78),

$$\sum_{t=s}^{s+T-1} f_0(\widehat{x}_t, \widehat{u}_t) \le \sum_{t=s}^{s+T-1} f(t, \widehat{x}, \widehat{u}) + M_0 + \epsilon/4$$

$$\le T f_0(\bar{x}, \bar{u}) + \delta T + \epsilon/4 + M_0 \le T f_0(\bar{x}, \bar{u}) + M_0 + 2. \tag{8.79}$$

Property (P2) and (8.79) imply that

$$\min\{\rho_E(\widehat{x}_i, \bar{x}) + \rho_F(\widehat{u}_i, \bar{u}) : i = s + 1, \ldots, s + T - 1\} \le \epsilon/4.$$

When combined with (8.76) and (8.77) this implies that

$$\min\{\rho_E(x_t, \bar{x}) + \rho_F(u_t, \bar{u}) : t = s + 1, \ldots, s + T - 1\} \le \epsilon.$$

Lemma 8.10 is proved.

Lemma 8.11 *For each natural number T,*

$$U^{f_*}_{\mathcal{M}_*, G_*}(0, T, \bar{x}, \bar{x}) = T f_0(\bar{x}, \bar{u}).$$

Proof Let T be a natural number. Clearly,

$$U^{f_*}_{\mathcal{M}_*, G_*}(0, T, \bar{x}, \bar{x}) \leq T f_0(\bar{x}, \bar{u}).$$

Assume that

$$(\{x_t\}_{t=0}^{T}, \{u_t\}_{t=0}^{T-1}) \in X(0, T, \mathcal{M}, G), \ x_0 = x_T = \bar{x}.$$

Set

$$u_T = \bar{u}.$$

For each integer $i > T$, define $x_i \in E, u_i \in F$ such that

$$x_{i+T} = x_i, \ u_{i+T} = u_i$$

for each integer $i \geq 0$. Clearly,

$$(\{x_t\}_{t=0}^{\infty}, \{u_t\}_{t=0}^{\infty}) \in X(0, \infty, \mathcal{M}_*, G_*).$$

By assumption (A2), for each natural number k,

$$T k f_0(\bar{x}, \bar{u}) + \bar{c} \leq \sum_{t=0}^{kT-1} f_0(x_t, u_t) = k \sum_{t=0}^{T-1} f_0(x_t, u_t)$$

for all natural numbers k. This implies that

$$\sum_{t=0}^{T-1} f_0(x_t, u_t) \geq T f_0(\bar{x}, \bar{u}).$$

Lemma 8.11 is proved.

Lemma 8.12 *Let $\epsilon > 0$. Then there exists $\delta \in (0, \bar{\lambda})$ such that for each natural number T and each $z_1, z_2 \in B_E(\bar{x}, \delta)$ the inequality*

$$|U^{f_*}_{\mathcal{M}_*, G_*}(0, T, z_1, z_2) - T f_0(\bar{x}, \bar{u})| \leq \epsilon$$

holds.

Proof In view the continuity of f at $(\bar{x}, \bar{u})$ there exists

$$\delta_0 \in (0, \min\{\bar{\lambda}, \ \epsilon\})$$

such that for each

$$(z, \xi) \in B_E(\bar{x}, \delta_0) \times B_F(\bar{u}, \delta_0)$$

we have

$$|f_0(z, \xi) - f_0(\bar{x}, \bar{u})| \leq \epsilon/8. \tag{8.80}$$

By (A1), there exists a positive number $\delta \in (0, \delta_0)$ such that the following property holds:

(i) for each $z_1, z_2 \in B_E(\bar{x}, \delta)$ there exists

$$u \in \mathcal{U}_0(z_1) \cap B(\bar{u}, \delta_0)$$

such that

$$z_2 \in G_0(z_1, u).$$

Assume that T is a natural number and that

$$z_1, z_2 \in B_E(\bar{x}, \delta). \tag{8.81}$$

First we show that

$$U^{f_*}_{\mathcal{M}_*, G_*}(0, T, z_1, z_2) \leq T f_0(\bar{x}, \bar{u}) + \epsilon.$$

There are two cases: $T = 1$; $T > 1$. Assume that $T = 1$. By (8.81), there exists

$$u \in \mathcal{U}_0(z_1) \cap B(\bar{u}, \delta_0) \tag{8.82}$$

such that

$$z_2 \in G_0(z_1, u). \tag{8.83}$$

In view of (8.82) and (8.83),

$$(\{z_1, z_2\}, \{u\}) \in X(0, 1, \mathcal{M}, G),$$

$$U^{f_*}_{\mathcal{M}_*, G_*}(0, 1, z_1, z_2) \leq f_0(z_1, u). \tag{8.84}$$

By (8.80)–(8.82),

$$f_0(z_1, z_2) \leq f_0(\bar{x}, \bar{u}) + \epsilon/8.$$

Together with (8.84) this implies that

$$U_{\mathcal{M}_*, G_*}^{f_*}(0, 1, z_1, z_2) \leq f_0(\bar{x}, \bar{u}) + \epsilon/8.$$

Assume that $T > 1$. Set

$$x_0 = z_1, \ \ x_T = z_2, \ \ x_t = \bar{x}, \ t \in \{1, \ldots, T-1\}. \tag{8.85}$$

By (8.81) and property (a), there exists

$$u_0 \in \mathcal{U}(z_1) \cap B(\bar{u}, \delta_0) \tag{8.86}$$

such that

$$\bar{x} \in G_0(z_1, u_0) \tag{8.87}$$

and

$$u_{T-1} \in \mathcal{U}(\bar{x}) \cap B(\bar{u}, \delta_0) \tag{8.88}$$

such that

$$z_2 \in G_0(\bar{x}, u_{T-1}). \tag{8.89}$$

Set

$$u_t = \bar{u}, \ t \in \{1, \ldots, T-1\} \setminus \{T-1\}.$$

Clearly,

$$(\{x_t\}_{t=0}^{T}, \{u_t\}_{t=0}^{T-1}) \in X(0, T, \mathcal{M}_*, G_*).$$

By (8.80), (8.81), (8.85)–(8.89),

$$U_{\mathcal{M}_*, G_*}^{f_*}(0, T, z_1, z_2) \leq \sum_{t=0}^{T-1} f_0(x_t, u_t)$$

$$\leq T f_0(\bar{x}, \bar{u}) + |f_0(x_0, u_0) - f_0(\bar{x}, \bar{u})| + |f_0(x_{T-1}, u_{T-1}) - f_0(\bar{x}, \bar{u})|$$

$$\leq T f_0(\bar{x}, \bar{u}) + \epsilon/4.$$

Thus

$$U_{\mathcal{M}_*,G_*}^{f_*}(0, T, z_1, z_2) \le T f_0(\bar{x}, \bar{u}) + \epsilon/4 \tag{8.90}$$

in both cases.

Assume that

$$(\{x_t\}_{t=0}^{T}, \{u_t\}_{t=0}^{T-1}) \in X(0, T, \mathcal{M}_*, G_*),$$

$$x_0 = z_1, \ x_T = z_2.$$

Set

$$\tilde{x}_0 = \bar{x}, \ \tilde{x}_t = x_{t-1}, \ t \in \{1, \ldots, T+1\}, \ \tilde{x}_{T+2} = \bar{x}, \tag{8.91}$$

$$\tilde{u}_t = u_{t-1}, \ t \in \{1, \ldots, T\}. \tag{8.92}$$

Property (a) and (8.81) imply that there exist

$$\tilde{u}_0 \in \mathcal{U}(\bar{x}) \cap B(\bar{u}, \delta_0) \tag{8.93}$$

such that

$$z_1 \in G_0(\bar{x}, \tilde{u}_0) \tag{8.94}$$

and

$$\tilde{u}_{T+1} \in \mathcal{U}(z_2) \cap B(\bar{u}, \delta_0) \tag{8.95}$$

such that

$$\bar{x} \in G_0(z_2, \tilde{u}_{T+1}). \tag{8.96}$$

In view of (8.91)–(8.96) and Lemma 8.11,

$$(\{\tilde{x}_t\}_{t=0}^{T+2}, \{\tilde{u}_t\}_{t=0}^{T+1}) \in X(0, T+2, \mathcal{M}_*, G_*),$$

$$(T+2) f_0(\bar{x}, \bar{u}) = U_{\mathcal{M}_*,G_*}^{f_*}(0, T+2, \bar{x}, \bar{u}) \le \sum_{t=0}^{T+1} f_0(\tilde{x}_t, \tilde{u}_t)$$

$$= f_0(\tilde{x}_0, \tilde{u}_0) + \sum_{t=0}^{T-1} f_0(x_t, u_t) + f_0(z_2, \tilde{u}_{T+1}). \tag{8.97}$$

By (8.80), (8.93) and (8.95),

$$|f_0(\tilde{x}_0, \tilde{u}_0) - f_0(\bar{x}, \bar{u})| \le \epsilon/8, \quad |f_0(z_2, \tilde{u}_{T+1}) - f_0(\bar{x}, \bar{u})| \le \epsilon/8. \tag{8.98}$$

It follows from (8.97) and (8.98) that

$$(T + 2)f_0(\bar{x}, \bar{u}) \le \sum_{t=0}^{T-1} f_0(x_t, u_t) + 2f_0(\bar{x}, \bar{u}) + \epsilon/4 + \epsilon/4,$$

$$\sum_{t=0}^{T-1} f_0(x_t, u_t) \ge T f_0(\bar{x}, \bar{u}) - \epsilon/2.$$

Since $(\{x_t\}_{t=0}^{T}, \{u_t\}_{t=0}^{T-1})$ is any element of $X(0, T, \mathcal{M}_*, G_*)$ we conclude that

$$T f_0(\bar{x}, \bar{u}) - \epsilon/2 \le U_{\mathcal{M}_*, G_*}^{f_*}(0, T, z_1, z_2).$$

When combined with (8.90) this inequality completes the proof of Lemma 8.12.

Lemma 8.13 *Let ϵ be a positive number. Then there exists $\delta \in (0, \bar{\lambda})$ such that for each natural number L there exists $\delta_0 \in (0, \delta)$ such that the following assertion holds.*

For each integer $s \ge 0$, each natural number $T \le L$, each $f \in \mathfrak{M}$ satisfying

$$\|f - f_*\| \le \delta_0, \tag{8.99}$$

each

$$(\mathcal{A}, \mathcal{U}, G) \in \mathcal{E}(\delta_0), \tag{8.100}$$

$$\mathcal{M} = graph(\mathcal{U}), \tag{8.101}$$

each

$$(\{x_t\}_{t=s}^{s+T}, \{u_t\}_{t=s}^{s+T-1}) \in X(s, s + T, \mathcal{M}, G), \tag{8.102}$$

satisfying

$$\rho_E(x_s, \bar{x}), \quad \rho_E(x_{s+T}, \bar{x}) \le \delta \tag{8.103}$$

and

$$\sum_{t=s}^{s+T-1} f(t, x_t, u_t) \le U_{\mathcal{M}, G}^{f}(s, s + T, x_s, x_{s+T}) + \delta \tag{8.104}$$

the inequalities

$$\rho_E(x_t, \bar{x}) \le \epsilon, \ t = s, \ldots, s + T,$$

$$\rho_F(u_t, \bar{u}) \le \epsilon, \ t = s + 1, \ldots, s + T - 1$$

hold.

Proof By Lemma 8.6, there exists a positive number

$$\gamma < \min\{\epsilon/8, \ \bar{\lambda}/4\}$$

such that the following property holds:

(P4)   for each integer $T \ge 1$ and each $(\{x_t\}_{t=0}^{T}, \{u_t\}_{t=0}^{T-1}) \in X(0, T, \mathcal{M}_*, G_*)$ which satisfies

$$\rho_E(x_0, \bar{x}), \ \rho_E(x_T, \bar{x}) \le 2\gamma,$$

$$\sum_{t=0}^{T-1} f_0(x_t, u_t) \le U_{\mathcal{M}_*, G_*}^{f_*}(0, T, x_0, x_T) + 2\gamma$$

we have $\rho_E(x_t, \bar{x}), \ \rho_F(u_t, \bar{x}) \le \epsilon/4$ holds for all $t = 0, \ldots, T - 1$.

Since f_0 is continuous at $(\bar{x}, \bar{u})$ there exists

$$\gamma_0 \in (0, \gamma/2)$$

such that

$$|f_0(\xi_1, \xi_2) - f_0(\bar{x}, \bar{u})| \le \gamma/4 \tag{8.105}$$

for each $\xi_1 \in B_E(\bar{x}, \gamma_0)$ and each $\xi_2 \in B_F(\bar{u}, \gamma_0)$.
  By (A1), there exists

$$\gamma_1 \in (0, \gamma_0/4)$$

such that the following property holds:
  (P5) for each $\xi_1, \xi_2 \in B(\bar{x}, \gamma_1)$ there exist

$$\eta \in \mathcal{U}_0(\xi_1) \cap B_F(\bar{u}, \gamma_0)$$

such that

$$\xi_2 \in G_0(\xi_1, \eta).$$

By Lemma 8.12, there exists $\delta \in (0, \gamma_1/4)$ such that the following property holds:

(P6)   for each natural number T and each $z_0, z_1 \in B_E(\bar{x}, 2\delta)$ the inequality

$$|U^{f_*}_{\mathcal{M}_*, G_*}(0, T, z_1, z_2) - T f_0(\bar{x}, \bar{u})| \leq \gamma/4$$

holds.

Let L be a natural number. By Lemma 8.9, there exists $\delta_1 \in (0, \delta)$ such that the following property holds:

(P7)   for each integer $s \geq 0$, each integer $T \in \{1, \ldots, L\}$, each $f \in \mathfrak{M}$ satisfying

$$\|f - f_*\| \leq \delta_1,$$

each

$$(\mathcal{A}, \mathcal{U}, G) \in \mathcal{E}(\delta_1), \ \ \mathcal{M} = \mathrm{graph}(\mathcal{U}),$$

each $(\{x_t\}_{t=s}^{s+T}, \{u_t\}_{t=s}^{s+T-1}) \in X(s, s+T, \mathcal{M}, G)$ there exists

$$(\{\widehat{x}_t\}_{t=s}^{s+T}, \{\widehat{u}_t\}_{t=s}^{s+T-1}) \in X(s, s+T, \mathcal{M}_*, G_*)$$

such that

$$\rho_E(x_t, \widehat{x}_t) \leq \delta/4, \ t = s, \ldots, s+T, \ \rho_F(u_t, \widehat{u}_t) \leq \delta/4,$$
$$t = s, \ldots, s+T-1 \tag{8.106}$$

and

$$\sum_{t=s}^{s+T-1} f_0(\widehat{x}_t, \widehat{u}_t) \leq \sum_{t=s}^{s+T-1} f(t, x_t, u_t) + \delta/4. \tag{8.107}$$

Set

$$\delta_0 = (2L)^{-1}\delta_1. \tag{8.108}$$

Assume that $s \geq 0$ is an integer, $T \in \{1, \ldots, L\}$, $f \in \mathfrak{M}$, (8.99)–(8.101) hold, $(\{x_t\}_{t=s}^{s+T}, \{u_t\}_{t=s}^{s+T-1}) \in X(s, s+T, \mathcal{M}, G)$ satisfies (8.103) and (8.104). By (8.99), (8.101), (8.108) and property (P7), there exists $(\{\widehat{x}_t\}_{t=s}^{s+T}, \{\widehat{u}_t\}_{t=s}^{s+T-1}) \in X(s, s+T, \mathcal{M}_*, G_*)$ satisfying (8.106), (8.107).

In view of (8.104) and (8.107),

$$\sum_{t=s}^{s+T-1} f_0(\widehat{x}_t, \widehat{u}_t) \le U^f_{\mathcal{M},G}(s, s+T, x_s, x_{s+T}) + \delta + \delta/4. \tag{8.109}$$

Equations (8.99), (8.108) and (8.109) imply that

$$\sum_{t=s}^{s+T-1} f_0(\widehat{x}_t, \widehat{u}_t) \le U^{f_*}_{\mathcal{M}_*,G_*}(s, s+T, x_s, x_{s+T}) + \delta + \delta/4 + \delta_1/2. \tag{8.110}$$

In view of (8.106), for $i = s, s+T$,

$$\rho_E(\widehat{x}_i, \bar{x}) \le \rho_E(\widehat{x}_i, x_i) + \rho_E(x_i, \bar{x}) \le \delta/4 + \delta.$$

Together with (8.110) and property (P4) this implies that

$$\rho_E(\widehat{x}_t, \bar{x}) \le \epsilon/4, \ \ t = s, \ldots, s+T,$$

$$\rho_F(\widehat{u}_t, \bar{u}) \le \epsilon/4, \ \ t = s, \ldots, s+T-1. \tag{8.111}$$

By (8.106) and (8.111),

$$\rho_E(x_t, \bar{x}) \le \epsilon, \ \ t = s, \ldots, s+T,$$

$$\rho_F(u_t, \bar{u}) \le \epsilon, \ \ t = s, \ldots, s+T-1.$$

This completes the proof of Lemma 8.13.

Lemma 8.14 *Let $\epsilon \in (0, \widehat{\lambda})$, $M_0 > 0$ and l_1, l_2 be natural numbers. Then there exists $\delta \in (0, \epsilon)$ and a natural number $L > l_1 + l_2$ such that for each integer $s \ge 0$, each natural number $T \ge L$, each $f \in \mathfrak{M}$ satisfying*

$$\|f - f_*\| \le \delta, \tag{8.112}$$

each

$$(\mathcal{A}, \mathcal{U}, G) \in \mathcal{E}(\delta), \ \ \mathcal{M} = graph(\mathcal{U}) \tag{8.113}$$

and each

$$(\{x_t\}_{t=s}^{s+T}, \{u_t\}_{t=s}^{s+T-1}) \in X(s, s+T, \mathcal{M}, G)$$

for which at least of the following conditions hold:

(a)

$$x_{T+s} \in Y(s + T - l_2, s + T, \mathcal{M}, G),$$

$$\sum_{t=s}^{s+T-1} f(t, x_t, u_t) \leq U_{\mathcal{M}, G}^f(s, s + T, x_s, x_{s+T}) + M_0;$$

(b)

$$\sum_{t=s}^{s+T-1} f(t, x_t, u_t) \leq U_{\mathcal{M}, G}^f(s, s + T, x_s) + M_0$$

and each integer $\tau \in [s, s + T - L]$ satisfying

$$x_\tau \in \bar{Y}(\tau, \tau + l_1, \mathcal{M}, G) \tag{8.114}$$

the inequality

$$\min\{\rho_E(x_t, \bar{x}) + \rho_F(u_t, \bar{u}) : \ t = \tau, \ldots, \tau + L - 1\} \leq \epsilon$$

holds.

Proof We may assume without loss of generality that

$$\epsilon < \bar{\lambda}.$$

By Lemma 8.10, there exists a natural number $L_0 \geq 4$ and $\delta \in (0, \epsilon)$ such that the following property holds:

(P8)   for each $f \in \mathfrak{M}$ satisfying $\| f - f_* \| \leq \delta$, each

$$(\mathcal{A}, \mathcal{U}, G) \in \mathcal{E}(\delta), \ \mathcal{M} = \text{graph}(\mathcal{U}),$$

each integer $s \geq 0$, each

$$(\{x_t\}_{t=s}^{s+L_0}, \{u_t\}_{t=s}^{s+L_0-1}) \in X(s, s + L_0, \mathcal{M}, G)$$

satisfying

$$\sum_{t=s}^{s+L_0-1} f(t, x_t, u_t) \leq \sum_{t=s}^{s+L_0-1} f(t, \bar{x}, \bar{u}) + 1$$

we have

$$\min\{\rho_E(x_t, \bar{x}) + \rho_F(u_t, \bar{u}) : t = s + 1, \ldots, s + L_0 - 1\} \le \epsilon.$$

Choose a natural number k_0 such that

$$k_0 > 4(\|f_*\| + 1)(L_0 + l_1 + l_2 + 1) + M_0 + 2. \tag{8.115}$$

Set

$$L = L_0 k_0. \tag{8.116}$$

Assume that $s \ge 0$, each natural number $T \ge L$, $f \in \mathfrak{M}$ satisfies (8.112), Eq. (8.113) holds,

$$(\{x_t\}_{t=s}^{s+T}, \{u_t\}_{t=s}^{s+T-1}) \in X(s, s + T, \mathcal{M}, G),$$

$$\mathcal{M} = \mathrm{graph}(\mathcal{U}),$$

at least of the conditions (a), (b) holds, $\tau \in [s, s + T - L]$ and (8.114) holds. In order to complete the proof of the lemmas it is sufficient to show that

$$\min\{\rho_E(x_t, \bar{x}) + \rho_F(u_t, \bar{u}) : t = \tau + 1, \ldots, \tau + L - 1\} \le \epsilon.$$

Assume the contrary. Then

$$\rho_E(x_t, \bar{x}) + \rho_F(u_t, \bar{u}) > \epsilon, \ t = \tau + 1, \ldots, \tau + L - 1. \tag{8.117}$$

There are two cases:

(1) there is an integer

$$S_0 \in [\tau + L, s + T] \setminus \{s + T\}$$

such that

$$\rho_E(x_{S_0}, \bar{x}) + \rho_F(u_{S_0}, \bar{x}) \le \epsilon; \tag{8.118}$$

(2)

$$\rho_E(x_t, \bar{x}) + \rho_F(u_t, \bar{u}) > \epsilon \tag{8.119}$$

for all integers $t \in [\tau + L, s + T] \setminus \{s + T\}$.

Assume that the case (1) holds. In view of (8.116), we may assume without loss of generality that

$$\rho_E(x_t, \bar{x}) + \rho_F(u_t, \bar{u}) > \epsilon, \ t \in \{\tau + L, \ldots, S_0\} \setminus \{S_0\}. \tag{8.120}$$

By (8.114), there exists

$$(\{\tilde{x}_t\}_{t=\tau}^{\tau+l_1}, \{\tilde{u}_t\}_{t=\tau}^{\tau+l_1-1}) \in X(\tau, \tau + l_1, \mathcal{M}, G)$$

such that

$$\tilde{x}_\tau = x_\tau, \ \tilde{x}_{\tau+l_1} = \bar{x}. \tag{8.121}$$

Set

$$\tilde{x}_t = \bar{x}, \ t \in \{\tau + l_1, \ldots, S_0 - 1\}, \tag{8.122}$$

$$\tilde{u}_t = \bar{u}, \ t \in \{\tau + l_1, \ldots, S_0 - 2\}. \tag{8.123}$$

In view of (A1), (B3), (8.118), (8.121) and the inequality $\epsilon < \widehat{\lambda}$, there exists an

$$\tilde{u}_{S_0-1} \in \mathcal{U}_0(\bar{x}) \cap B(\bar{u}, \bar{\lambda})$$

such that

$$x_{S_0} \in G_0(\bar{x}, \tilde{u}_{S_0-1}).$$

By (8.121)–(8.123) and the relations above,

$$(\{\tilde{x}_t\}_{t=\tau}^{S_0}, \{\tilde{u}_t\}_{t=\tau}^{S_0-1}) \in X(\tau, S_0, \mathcal{M}, G).$$

Conditions (a), (b), (8.121)–(8.123) and the relations above imply that

$$\sum_{t=\tau}^{S_0-1} f(t, x_t, u_t) \le \sum_{t=\tau}^{S_0-1} f(t, \tilde{x}_t, \tilde{u}_t) + M_0$$

$$\le \sum_{t=\tau_0}^{S_0-1} f(t, \bar{x}, \bar{u}) + 2l_1 \|f\| + 2\|f\| + M_0. \tag{8.124}$$

There exists an integer $k \ge 0$ such that

$$S_0 - \tau \in (kL_0, (k+1)L_0]. \tag{8.125}$$

Since we consider case (1) it follows from (8.116) and (8.125) that

$$k \ge k_0. \tag{8.126}$$

It follows from (8.112), (8.113), (8.117), (8.120), (8.125) and property (P8) that the following property holds:

(P9)   for each integer $j \in [0, k-2]$,

$$\sum_{t=\tau+jL_0}^{\tau+(j+1)L_0-1} f(t, x_t, u_t) > \sum_{t=\tau+jL_0}^{\tau+(j+1)L_0-1} f(t, \bar{x}, \bar{x}) + 1.$$

Property (P9) and (8.112), (8.124) and (8.126) imply that

$$M_0 + 2(l_1 + 1)(\|f_*\| + 1) \geq \sum_{t=\tau}^{S_0-1} f(t, x_t, u_t) - \sum_{t=\tau}^{S_0-1} f(t, \bar{x}, \bar{x})$$

$$\geq \sum_{j=0}^{k-2} \left[\sum_{t=\tau+jL_0}^{\tau+(j+1)L_0-1} (f(t, x_t, u_t) - f(t, \bar{x}, \bar{x})) \right]$$

$$- 4L_0(\|f_*\| + 1)$$

$$\geq k - 8L_0(\|f_*\| + 1) \geq k_0 - 8L_0(\|f_*\| + 1)$$

and

$$k_0 \leq 4(\|f_*\| + 1)(L_0 + l_1 + 1) + M_0 + 2.$$

This contradicts (8.115). The contradiction we have reached proves that case (1) does not hold. Therefore case (2) holds and (8.119) is true.

In view of (8.114), there exists

$$(\{\tilde{x}_t\}_{t=\tau}^{\tau+l_1}, \{\tilde{u}_t\}_{t=\tau}^{\tau+l_1-1}) \in X(\tau, \tau + l_1, \mathcal{M}, G)$$

such that

$$\tilde{x}_\tau = x_\tau, \ \tilde{x}_{\tau+l_1} = \bar{x}. \tag{8.127}$$

In the case of condition (b) set

$$\tilde{x}_t = \bar{x}, \ t \in \{\tau + l_1, \ldots, s + T\} \setminus \{\tau + l_1\},$$

$$\tilde{u}_t = \bar{u}, \ t \in \{\tau + l_1, \ldots, s + T\} \setminus \{s + T\}. \tag{8.128}$$

By (8.127) and (8.128),

$$(\{\tilde{x}_t\}_{t=\tau}^{T+s}, \{\tilde{u}_t\}_{t=\tau}^{T+s-1}) \in X(\tau, T + s, \mathcal{M}, G).$$

In the case (a) there exists

$$(\{\tilde{x}_t\}_{t=T-l_2+s}^{T+s}, \{\tilde{u}_t\}_{t=T-l_2+s}^{T-1+s}) \in X(T-l_2+s, T+s, \mathcal{M}, G)$$

satisfying

$$\tilde{x}_{T-l_2+s} = \bar{x}, \ \tilde{x}_{T+s} = x_{T+s}. \tag{8.129}$$

Set

$$\tilde{x}_t = \bar{x}, \ t \in \{\tau + l_1, \ldots, s + T - l_2\} \setminus \{\tau + l_1, T - l_2 + s\}, \tag{8.130}$$

$$\tilde{u}_t = \bar{u}, \ t \in \{\tau + l_1, \ldots, s + T - l_2\} \setminus \{s + T - l_2\}. \tag{8.131}$$

It follows from (8.127)–(8.131) that in both cases

$$(\{\tilde{x}_t\}_{t=\tau}^{s+T}, \{\tilde{u}_t\}_{t=\tau}^{s+T-1}) \in X(\tau, T+s, \mathcal{M}, G),$$

$$\sum_{t=\tau}^{s+T-1} f(t, x_t, u_t) \le \sum_{t=\tau}^{s+T-1} f(t, \tilde{x}_t, \tilde{u}_t) + M_0. \tag{8.132}$$

By (8.128), (8.130), (8.131), in both cases

$$\tilde{x}_t = \bar{x}, \ \tilde{u}_t = \bar{u}, \ t \in \{\tau + l_1, \ldots, s + T - l_2 - 1\}. \tag{8.133}$$

Equations (8.112), (8.132) and (8.133) imply that

$$\sum_{t=\tau}^{s+T-1} f(t, x_t, u_t) \le \sum_{t=\tau}^{s+T-1} f(t, \tilde{x}_t, \tilde{u}_t) + M_0$$

$$\le \sum_{t=\tau}^{s+T-1} f(t, \bar{x}, \bar{u}) + 2(l_2 + l_1)\|f\| + M_0$$

$$\le \sum_{t=\tau}^{s+T-1} f(t, \bar{x}, \bar{u}) + 2(l_2 + l_1)(\|f_*\| + 1) + M_0. \tag{8.134}$$

There exists a natural number k such that

$$T + s - \tau \in [kL_0, (k+1)L_0). \tag{8.135}$$

By (8.116) and (8.135),

$$k \ge k_0. \tag{8.136}$$

Since we have case (2) it follows from property (P8), (8.112), (8.115), (8.119) and (8.135) that for each integer $j \in [0, k-2]$,

$$\sum_{t=\tau+jL_0}^{\tau+(j+1)L_0-1} f(t, x_t, u_t) > \sum_{t=\tau+jL_0}^{\tau+(j+1)L_0-1} f(t, \bar{x}, \bar{x}) - 1. \tag{8.137}$$

By (8.134), (8.136) and (8.137),

$$M_0 + 2(l_1 + l_2)(\|f_*\| + 1) \geq \sum_{t=\tau}^{s+T-1} f(t, x_t, u_t) - \sum_{t=\tau}^{s+T-1} f(t, \bar{x}, \bar{u})$$

$$\geq \sum_{j=0}^{k-2} \sum_{t=\tau+jL_0}^{\tau+(j+1)L_0-1} (f(t, x_t, u_t) - f(t, \bar{x}, \bar{u}))$$

$$- 4L_0\|f\| \geq k - 1 - 4L_0(\|f_*\| + 1) \geq k_0 - 4L_0(\|f_*\| + 1) - 1$$

and

$$k_0 \leq 2(\|f_*\| + 1)(2L_0 + l_1 + l_2 + 1) + M_0 + 1.$$

This contradicts (8.145). The contradiction we have reached completes the proof of Lemma 8.14.

8.4   Proof of Theorems 8.3

By Lemma 8.13, there exists $\gamma \in (0, \widehat{\lambda})$ such that the following property holds:

(P10)   for each natural number L there exists $\gamma_L \in (0, \gamma)$ such that for each natural number $T \leq L$, each $f \in \mathfrak{M}$ satisfying

$$\|f - f_*\| \leq \gamma_L,$$

each

$$(\mathcal{A}, \mathcal{U}, G) \in \mathcal{E}(\gamma_L), \quad \mathcal{M} = \text{graph}(\mathcal{U}),$$

each integer $s \geq 0$, each

$$(\{x_t\}_{t=s}^{s+T}, \{u_t\}_{t=s}^{s+T-1}) \in X(s, s+T, \mathcal{M}, G)$$

satisfying

$$\rho_E(x_s, \bar{x}), \ \rho_E(x_{s+T}, \bar{x}) \leq \gamma,$$

and

$$\sum_{t=s}^{s+T-1} f(t, x_t, u_t) \leq U_{\mathcal{M},G}^f(s, s+T, x_s, x_{s+T}) + \gamma$$

we have

$$\rho_E(x_t, \bar{x}) \leq \epsilon, \ t = s, \ldots, s+T,$$

$$\rho_F(u_t, \bar{u}) \leq \epsilon, \ t = s+1, \ldots, s+T-1.$$

By Lemma 8.14 (with $\epsilon = \gamma$ and $M_0 = 1$), there exist

$$\tilde{\gamma} \in (0, \gamma)$$

and a natural number

$$L > l_1 + l_2 \tag{8.138}$$

such that the following properties hold:

(P11) for each integer $s \geq 0$, each natural number $T \geq L$, each $f \in \mathfrak{M}$ satisfying

$$\|f - f_*\| \leq \tilde{\gamma},$$

each

$$(\mathcal{A}, \mathcal{U}, G) \in \mathcal{E}(\tilde{\gamma}), \ \mathcal{M} = \mathrm{graph}(\mathcal{U}),$$

each

$$(\{x_t\}_{t=s}^{s+T}, \{u_t\}_{t=s}^{s+T-1}) \in X(s, s+T, \mathcal{M}, G)$$

for which at least of the following conditions hold:

(a)

$$x_{T+s} \in Y(s+T-l_2, s+T, \mathcal{M}, G),$$

$$\sum_{t=s}^{s+T-1} f(t, x_t, u_t) \leq U_{\mathcal{M},G}^f(s, s+T, x_s, x_{s+T}) + 1;$$

(b)

$$\sum_{t=s}^{s+T-1} f(t, x_t, u_t) \leq U^f_{\mathcal{M},G}(s, s+T, x_s) + 1$$

and each integer $\tau \in [s, s+T-L]$ satisfying

$$x_\tau \in \bar{Y}(\tau, \tau + l_1, \mathcal{M}, G)$$

the inequality

$$\min\{\rho_E(x_t, \bar{x}) + \rho_F(u_t, \bar{u}) : \ t = \tau + 1, \ldots, \tau + L - 1\} \leq \gamma$$

holds.

By (P10) there exists $\delta \in (0, \tilde{\gamma})$ such that the following property holds:

(P12)   for each natural number $T \leq 2L + 4$, each each $f \in \mathfrak{M}$ satisfying

$$\|f - f_*\| \leq \delta,$$

each

$$(\mathcal{A}, \mathcal{U}, G) \in \mathcal{E}(\delta), \ \mathcal{M} = \mathrm{graph}(\mathcal{U}),$$

each integer $s \geq 0$, each

$$(\{x_t\}_{t=s}^{s+T}, \{u_t\}_{t=s}^{s+T-1}) \in X(s, s+T, \mathcal{M}, G)$$

satisfying

$$\rho_E(x_s, \bar{x}), \ \rho_E(x_{s+T}, \bar{x}) \leq \gamma,$$

and

$$\sum_{t=s}^{s+T-1} f(t, x_t, u_t) \leq U^f_{\mathcal{M},G}(s, s+T, x_s, x_{s+T}) + \gamma$$

we have

$$\rho_E(x_t, \bar{x}) \leq \epsilon, \ t = s, \ldots, s+T,$$

$$\rho_F(u_t, \bar{u}) \leq \epsilon, \ t = s+1, \ldots, s+T-1.$$

Assume that integer $T > 2L$, $f \in \mathfrak{M}$ satisfies

$$\|f - f_*\| \le \delta, \ (\mathcal{A}, \mathcal{U}, G) \in \mathcal{E}(\delta), \tag{8.139}$$

$$\mathcal{M} = \mathrm{graph}(\mathcal{U}),$$

$(\{x_t\}_{t=0}^{T}, \{u_t\}_{t=0}^{T-1}) \in X(0, T, \mathcal{M}, G)$ satisfies

$$x_0 \in \bar{Y}(0, l_1, \mathcal{M}, G) \tag{8.140}$$

and at least one of the following conditions holds:

(a)

$$x_T \in Y(T - l_2, T, \mathcal{M}, G),$$

$$\sum_{t=0}^{T-1} f(t, x_t, u_t) \le U_{\mathcal{M},G}^{f}(0, T, x_0, x_T) + \delta;$$

(b)

$$\sum_{t=0}^{T-1} f(t, x_t, u_t) \le U_{\mathcal{M},G}^{f}(0, T, x_0) + \delta.$$

By (8.139), (8.140), the inequality $\gamma < \bar{\lambda}$, applying by induction property (P11) we obtain a finite sequence of integers S_i, $i = 0, \ldots, q$ such that

$$0 \le S_0 \le L, \ T - L < S_q \le T,$$

$$1 \le S_{i+1} - S_i \le L, \ i = 0, \ldots, q - 1,$$

$$\rho_E(x_{S_i}, \bar{x}) \le \gamma, \ i = 0, \ldots, q. \tag{8.141}$$

If $\rho_E(x_0, \bar{x}) \le \delta$, we may assume that $S_0 = 0$ and if $\rho_E(x_T, \bar{x}) \le \delta$, we may assume that $S_q = T$. Set

$$\tau_1 = S_0, \ \tau_2 = S_q.$$

Let an integer $t \in [\tau_1, \tau_2]$. Then there exists an integer $i \in [0, q - 1]$ such that

$$t \in [S_i, S_{i+1}].$$

Together with (8.141) and (P12) this implies that

$$\rho_E(x_t, \bar{x}) \le \epsilon, \quad \rho_F(u_t, \bar{u}) \le \epsilon.$$

This completes the proof of Theorems 8.3.

8.5   Proof of Theorem 8.4

By Lemma 8.13, there exists $\gamma \in (0, \widehat{\lambda})$ such that the following property holds:

(P13)   for each natural number L there exists $\gamma_L \in (0, \gamma)$ such that for each natural number $T \le L$, each $f \in \mathfrak{M}$ satisfying

$$\| f - f_* \| \le \gamma_L,$$

each

$$(\mathcal{A}, \mathcal{U}, G) \in \mathcal{E}(\gamma_L), \quad \mathcal{M} = \text{graph}(\mathcal{U}),$$

each integer $s \ge 0$, each

$$(\{x_t\}_{t=s}^{s+T}, \{u_t\}_{t=s}^{s+T-1}) \in X(s, s+T, \mathcal{M}, G)$$

satisfying

$$\rho_E(x_s, \bar{x}), \quad \rho_E(x_{s+T}, \bar{x}) \le \gamma,$$

and

$$\sum_{t=s}^{s+T-1} f(t, x_t, u_t) \le U_{\mathcal{M},G}^f(s, s+T, x_s, x_{s+T}) + \gamma,$$

we have

$$\rho_E(x_t, \bar{x}) \le \epsilon, \ t = s, \dots, s+T-1,$$

$$\rho_F(u_t, \bar{u}) \le \epsilon, \ t = s+1, \dots, s+T-1.$$

By Lemma 8.14 (with $\epsilon = \gamma$ and $M_0 = M+1$), there exist

$$\tilde{\gamma} \in (0, \gamma)$$

and a natural number

$$L_0 > l_1 + l_2$$

such that the following properties hold:

(P14)　for each integer $s \geq 0$, each natural number $T \geq L_0$, each $f \in \mathfrak{M}$ satisfying

$$\|f - f_*\| \leq \tilde{\gamma},$$

each

$$(\mathcal{A}, \mathcal{U}, G) \in \mathcal{E}(\tilde{\gamma}), \quad \mathcal{M} = \mathrm{graph}(\mathcal{U}),$$

each

$$(\{x_t\}_{t=s}^{s+T}, \{u_t\}_{t=s}^{s+T-1}) \in X(s, s+T, \mathcal{M}, G)$$

for which at least of the following conditions hold:

$$x_{T+s} \in Y(s+T-l_2, s+T, \mathcal{M}, G),$$

$$\sum_{t=s}^{s+T-1} f(t, x_t, u_t) \leq U_{\mathcal{M},G}^{f}(s, s+T, x_s, x_{s+T}) + M + 1;$$

$$\sum_{t=s}^{s+T-1} f(t, x_t, u_t) \leq U_{\mathcal{M},G}^{f}(s, s+T, x_s) + M + 1$$

and each integer $\tau \in [s, s+T-L_0]$ satisfying

$$x_\tau \in \bar{Y}(\tau, \tau+l_1, \mathcal{M}, G)$$

the inequality

$$\min\{\rho_E(x_t, \bar{x}) + \rho_F(u_t, \bar{x}) : \ t = \tau+1, \ldots, \tau+L_0-1\} \leq \gamma$$

holds.

By (P13) there exists $\delta \in (0, \tilde{\gamma})$ such that the following property holds:

(P15)　for each natural number $T \leq 2L_0 + 4$, each $f \in \mathfrak{M}$ satisfying

$$\|f - f_*\| \leq \delta,$$

each

$$(\mathcal{A}, \mathcal{U}, G) \in \mathcal{E}(\delta), \ \ \mathcal{M} = \mathrm{graph}(\mathcal{U}),$$

each integer $s \geq 0$, each

$$(\{x_t\}_{t=s}^{s+T}, \{u_t\}_{t=s}^{s+T-1}) \in X(s, s+T, \mathcal{M}, G)$$

satisfying

$$\rho_E(x_s, \bar{x}), \ \ \rho_E(x_{s+T}, \bar{x}) \leq \gamma,$$

and

$$\sum_{t=s}^{s+T-1} f(t, x_t, u_t) \leq U_{\mathcal{M},G}^f(s, s+T, x_s, x_{s+T}) + \gamma$$

we have

$$\rho_E(x_t, \bar{x}) \leq \epsilon, \ \ t = s, \ldots, s+T-1,$$

$$\rho_F(u_t, \bar{u}) \leq \epsilon, \ \ t = s, \ldots, s+T-1.$$

Choose a natural number

$$L > (4 + \gamma^{-1}(1 + M))(L_0 + 1). \tag{8.142}$$

Assume that integer $T > L$, $f \in \mathfrak{M}$ satisfies

$$\|f - f_*\| \leq \delta, \tag{8.143}$$

$$(\mathcal{A}, \mathcal{U}, G) \in \mathcal{E}(\delta),$$

$$\mathcal{M} = \mathrm{graph}(\mathcal{U},$$

$$(\{x_t\}_{t=0}^{T}, \{u_t\}_{t=0}^{T-1}) \in X(0, T, \mathcal{M}, G), \ x_0 \in \bar{Y}(0, l_1, \mathcal{M}, G) \tag{8.144}$$

and that at least one of the following conditions holds:

(a)

$$x_T \in Y(T - l_2, T, \mathcal{M}, G),$$

$$\sum_{t=0}^{T-1} f(t, x_t, u_t) \le U_{\mathcal{M},G}^{f}(0, T, x_0, x_T) + M;$$

(b)

$$\sum_{t=0}^{T-1} f(t, x_t, u_t) \le U_{\mathcal{M},G}^{f}(0, T, x_0) + M.$$

By (8.142)–(8.144), the inequality $\delta < \tilde{\gamma} < \widehat{\lambda}$, (a), (b), applying by induction property (P14) we obtain a finite sequence of integers S_i, $i = 0, \ldots, q$ such that

$$0 \le S_0 \le L_0, \ \ T - L_0 < S_q \le T, \tag{8.145}$$

$$1 \le S_{i+1} - S_i \le L, \ \ i = 0, \ldots, q - 1, \tag{8.146}$$

$$\rho_E(x_{S_i}, \bar{x}) \le \gamma, \ \ i = 0, \ldots, q. \tag{8.147}$$

Properties (a) and (b) imply that

$$M \ge \sum_{t=0}^{T-1} f(t, x_t, u_t) - U_{\mathcal{M},G}^{f}(0, T, x_0, x_T)$$

$$\ge \sum_{i=0}^{q-1} \Big(\sum_{t=S_i}^{S_{i+1}-1} f(t, x_t, u_t) - U_{\mathcal{M},G}^{f}(S_i, S_{i+1}, T, x_{S_i}, x_{S_{i+1}}) \Big). \tag{8.148}$$

Set

$$E = \{i \in \{0, \ldots, q - 1\} :$$

$$\sum_{t=S_i}^{S_{i+1}-1} f(t, x_t, u_t) - U_{\mathcal{M},G}^{f}(S_i, S_{i+1}, T, x_{S_i}, x_{S_{i+1}}) > \gamma \}. \tag{8.149}$$

By (8.148) and (8.149),

$$M \ge \gamma \, \mathrm{Card}(E)$$

and

$$\mathrm{Card}(E) \le \gamma^{-1} M. \tag{8.150}$$

Let

$$j \in \{0, \ldots, q-1\} \setminus E. \tag{8.151}$$

In view of (8.149) and (8.151),

$$\sum_{t=S_j}^{S_{j+1}-1} f(t, x_t, u_t) \le U^f_{\mathcal{M},G}(S_j, S_{j+1}, T, x_{S_j}, x_{S_{j+1}}) + \gamma. \tag{8.152}$$

Property (P15) and (8.143)–(8.145), (8.147), (8.152) imply that for $t = S_j, \ldots, S_{j+1} - 1$,

$$\rho_E(x_t, \bar{x}), \ \rho_F(u_t, \bar{u}) \le \epsilon. \tag{8.153}$$

It follows from (8.142), (8.145)–(8.147), (8.149)–(8.151), (8.153) that

$$\{t \in \{0, \ldots, T-1\} : \ \max\{\rho_E(x_t, \bar{x}), \ \rho_F(u_t, \bar{u})\} > \epsilon\}$$

$$\subset \cup\{\{S_i, \ldots, S_{i+1}\} : \ i \in E\} \cup \{0, \ldots, S_0\} \cup \{S_q, \ldots, T\},$$

$$\mathrm{Card}(\{t \in \{0, \ldots, T\} : \ \max\{\rho_E(x_t, \bar{x}), \ \rho_F(u_t, \bar{u})\} > \epsilon\})$$

$$\le (L_0 + 1)(\mathrm{Card}(E) + 2) \le (L_0 + 1)(2 + M\gamma^{-1}) < L.$$

This completes the proof of Theorem 8.4.

Chapter 9
Conclusions

The book presents our research on the turnpike phenomenon for Markov decision processes (MDPs). In contains two parts. In the first part we study MDPs themselves while in the second part we consider a general optimal control problem such that the deterministic optimal control problem associated with our MDP is its particular case. It should be mentioned that the first research on the turnpike theory for finite-state MDPs was done in [89]. The starting point of our research was our joint work with A. Leizarowitz [68]. The book contains a number of new results on properties of approximate solutions (policies) which are independent of the length of the interval, for all sufficiently large intervals. In the book we discuss infinite horizon discrete-time optimal control of MDPs with finite state spaces and compact action sets and employ the long-run expected average cost criterion. We study the uniqueness and stability of minimizing Markov actions and show that in a set of MDPs equipped with a natural complete metric subset of MDPs with essentially unique and stable minimizing Markov actions contains a countable intersection of open everywhere dense sets. Therefore, the property of having essentially unique and stable minimizing Markov actions is generic for this class of MDPs. Assuming the uniqueness of minimizing Markov actions we establish the existence of an overtaking optimal policy and the asymptotic turnpike property for good programs on infinite horizon. We also establish the weak turnpike property for approximate optimal programs on finite horizon. The final chapter of the first part of the book (Chap. 6) is devoted to the study of turnpike properties for MDPs with perturbations. We assume the uniqueness of minimizing Markov actions and show that the asymptotic turnpike property and the weak turnpike property are stable under the perturbations of cost functions. In the first part of the book (Chaps. 2–6) the study of our stochastic optimal control problem is reduced to the study of the corresponding deterministic optimal control problem which is a particular case of a general optimal control system which is studied in Chaps. 7 and 8, the second part of the book. Turnpike properties for deterministic optimal control systems are well-known in the literature but they are related to states of the system. But for our MDPs

A. J. Zaslavski, *Turnpike Phenomenon for Markov Decision Processes*,
SpringerBriefs in Mathematics, https://doi.org/10.1007/978-3-032-00854-1_9

and the related deterministic optimal control system the turnpike phenomenon holds both for its states and its controls. Therefore the existing turnpike results cannot be applied to our situation and new turnpike results are needed which are presented in the second part of the book, where we establish the strong turnpike property which is deduced from the asymptotic turnpike property and some controllability conditions.

References

1. Andrilli, S., Hecker, D.: Elementary Linear Algebra. Elsevier, Amsterdam (2010)
2. Arapostathis, A., Biswas, A.: Infinite horizon risk-sensitive control of diffusions without any blanket stability assumptions. Stochastic Process. Appl. **128**, 1485–1524 (2018). Infinite horizon stochastic control
3. Arapostathis, A., Biswas, A.: Risk-sensitive control for a class of diffusions with jumps. Ann. Appl. Probab. **32**, 4106–4142 (2022)
4. Arapostathis, A., Borkar, V., Fernandez-Gaucherand, E., Ghosh, M., Marcus, S.: Discrete-time controlled Markov processes with average cost criterion: a survey. SIAM J. Control Optim. **31**, 282–344 (1993)
5. Arapostathis, A., Biswas, A., Saha, S.: Strict monotonicity of principal eigenvalues of elliptic operators in R d and risk-sensitive control. J. Math. Pures Appl. **124**, 169–219 (2019)
6. Arkin, V.I., Evstigneev, I.V.: Stochastic Models of Control and Economic Dynamics. Academic Press, London (1987)
7. Aseev, S.M., Veliov, V.M.: Maximum principle for infinite-horizon optimal control problems with dominating discount. Dyn. Contin. Discrete Impuls. Syst. B **19**, 43–63 (2012)
8. Aseev, S.M., Krastanov, M.I., Veliov, V.M.: Optimality conditions for discrete-time optimal control on infinite horizon. Pure Appl. Funct. Anal. **2**, 395–409 (2017)
9. Aubin, J.P., Ekeland, I.: Applied Nonlinear Analysis. Wiley Interscience, New York (1984)
10. Aubry, S., Le Daeron, P.Y.: The discrete Frenkel-Kontorova model and its extensions I. Physica D **8**, 381–422 (1983)
11. Bachir, M., Blot, J.: Infinite dimensional infinite-horizon Pontryagin principles for discrete-time problems. Set-Valued Var. Anal. **23**, 43–54 (2015)
12. Bachir, M., Blot, J.: Infinite dimensional multipliers and Pontryagin principles for discrete-time problems. Pure Appl. Funct. Anal. **2**, 411–426 (2017)
13. Bather, J.: Optimal decision procedures for finite Markov chains, I. Adv. Appl. Probab. **5**, 328–339 (1973)
14. Bather, J.: Optimal decision procedures for finite Markov chains, II. Adv. Appl. Probab. **5**, 521–540 (1973)
15. Bather, J.: Optimal decision procedures for finite Markov chains, III. Adv. Appl. Probab. **5**, 541–553 (1973)
16. Baumeister, J., Leitao, A., Silva, G.N.: On the value function for nonautonomous optimal control problem with infinite horizon. Syst. Control Lett. **56**, 188–196 (2007)
17. Biswas, A., Pradhan, S.: Ergodic risk-sensitive control of Markov processes on countable state space revisited. ESAIM Control Optim. Calc. Var. **28**, 50 pp. (2022)

18. Blot, J.: Infinite-horizon Pontryagin principles without invertibility. J. Nonlinear Convex Anal. **10**, 177–189 (2009)
19. Blot, J., Hayek, N.: Sufficient conditions for infinite-horizon calculus of variations problems. ESAIM Control Optim. Calc. Var. **5**, 279–292 (2000)
20. Blot, J., Hayek, N.: Infinite-Horizon Optimal Control in the Discrete-Time Framework. SpringerBriefs in Optimization. Springer, New York (2014)
21. Borkar, V.S.: On minimum cost per unit time control of Markov chains. SIAM J. Control Optim. **22**, 965–984 (1984)
22. Borkar, V.S.: Control of Markov chains with long-run average cost criterion: the dynamic programming equations. SIAM J. Control Optim. **27**, 642–657 (1989)
23. Borkar, V.S., Miclo, L.: On the fastest finite Markov processes. J. Math. Anal. Appl. **481**, 43 pp. (2020)
24. Carlson, D.A.: The existence of catching-up optimal solutions for a class of infinite horizon optimal control problems with time delay. SIAM J. Control Optim. **28**, 402–422 (1990)
25. Carlson, D.A., Jabrane, A., Haurie, A.: Existence of overtaking solutions to infinite dimensional control problems on unbounded time intervals. SIAM J. Control Optim. **25**, 1517–1541 (1987)
26. Carlson, D.A., Haurie, A., Leizarowitz, A.: Infinite Horizon Optimal Control. Springer-Verlag, Berlin (1991)
27. Damm, T., Grune, L., Stieler, M., Worthmann, K.: An exponential turnpike theorem for dissipative discrete time optimal control problems. SIAM J. Control Optim. **52**, 1935–1957 (2014)
28. Dufour, F., Prieto-Rumeau, T.: Approximation of average cost Markov decision processes using empirical distributions and concentration inequalities. Stochastics **87**, 273–307 (2015)
29. Dufour, F., Prieto-Rumeau, T.: Absorbing Markov decision processes. ESAIM Control Optim. Calc. Var. **30**, 18 pp. (2024)
30. Federgruen, A., Schweitzer, J.P.: A fixed-point approach to undiscounted Markov renewal programs. SIAM J. Algebr. Discrete Methods **5**, 539–550 (1984)
31. Federgruen, A., Schweitzer, J.P., Tijms, H.C.: Denumerable undiscounted semi-Markov decision processes with unbounded rewards. Math. Oper. Res. **8**, 298–313 (1983)
32. Feinberg, E.A.: On controlled finite state Markov processes with compact control sets. Theor. Probab. Appl. **20**, 856–861 (1975)
33. Feinberg, E.A., Piunovskiy, A.: Nonatomic total rewards Markov decision processes with multiple criteria. J. Math. Anal. Appl. **273**, 93–111 (2002)
34. Feinberg, E.A., Piunovskiy, A.: Sufficiency of deterministic policies for atomless discounted and uniformly absorbing MDPs with multiple criteria. SIAM J. Control Optim. **57**, 163–191 (2019)
35. Gaitsgory, V., Parkinson, A., Shvartsman, I.: Linear programming formulations of deterministic infinite horizon optimal control problems in discrete time. Discrete Contin. Dyn. Syst. B **22**, 3821–3838 (2017)
36. Gale, D.: On optimal development in a multi-sector economy. Rev. Econ. Stud. **34**, 1–18 (1967)
37. Glizer, V.Y., Kelis, O.: Upper value of a singular infinite horizon zero-sum linear-quadratic differential game. Pure Appl. Funct. Anal. **2**, 511–534 (2017)
38. Grune, L., Guglielmi, R.: Turnpike properties and strict dissipativity for discrete time linear quadratic optimal control problems. SIAM J. Control Optim. **56**, 1282–1302 (2018)
39. Gugat, M.: A turnpike result for convex hyperbolic optimal boundary control problems. Pure Appl. Funct. Anal. **4**, 849–866 (2019)
40. Gugat, M., Hante, F.M.: On the turnpike phenomenon for optimal boundary control problems with hyperbolic systems. SIAM J. Control Optim. **57**, 264–289 (2019)
41. Gugat, M., Trelat, E., Zuazua, E.: Optimal Neumann control for the 1D wave equation: finite horizon, infinite horizon, boundary tracking terms and the turnpike property. Syst. Control Lett. **90**, 61–70 (2016)

42. Guo, X., Hernandez-Lerma, O.: Zero-sum continuous-time Markov games with unbounded transition and discounted payoff rates. Bernoulli **11**, 1009–1029 (2005)
43. Guo, X., Hernandez-Lerma, O.: Continuous-Time Markov Decision Processes. Stoch. Model. Appl. Probab., vol. 62. Springer, Berlin (2009)
44. Guo, X., Piunovskiy, A.: Discounted continuous-time Markov decision processes with constraints: unbounded transition and loss rates. Math. Oper. Res. **36**, 105–132 (2011)
45. Guo, X., Hernndez-del-Valle, A., Hernndez-Lerma, O.: Nonstationary discrete-time deterministic and stochastic control systems with infinite horizon. Int. J. Control **83**, 1751–1757 (2010)
46. Hayek, N.: Infinite horizon multiobjective optimal control problems in the discrete time case. Optimization **60**, 509–529 (2011)
47. Hernndez-Lerma, O., Hernndez-Hernndez, D.: Discounted cost Markov decision processes on Borel spaces: the linear programming formulation. J. Math. Anal. Appl. **183**, 335–351 (1994)
48. Hernndez-Lerma, O., Lasserre, J.B.: Discrete-Time Markov Control Processes. Springer, New York (1996)
49. Hernandez-Lerma, O., Lasserre, J.B.: Zero-sum stochastic games in Borel spaces: average payoff criteria. SIAM J. Control Optim. **39**, 1520–1539 (2001)
50. Hernndez-Lerma, O., Lasserre, J.B.: Markov Chains and Invariant Probabilities. Progress in Mathematics, vol. 211. Birkhuser Verlag, Basel (2003)
51. Hernndez-Lerma, O., Gonzlez-Hernndez, J., Lpez-Martnez Raquiel, R.: Constrained average cost Markov control processes in Borel spaces. SIAM J. Control Optim. **42**, 442–468 (2003)
52. Hinderer, K.: Foundation of Non-stationary Dynamic Programming with Discrete-Time Parameter. Lecture Notes in Operations Research, vol. 33. Springer-Verlag, New York (1970)
53. Hordijk, A.: Dynamic Programming and Markov Potential Theory. Tracts, vol. 51. Math. Centre., Amsterdam (1974)
54. Hordijk, A., Puterman, M.L.: On the convergence of policy iteration in undiscounted finite state Markov processes: the unichain case. Math. Oper. Res. **12**, 163–176 (1987)
55. Jasso-Fuentes, H., Hernandez-Lerma, O.: Characterizations of overtaking optimality for controlled diffusion processes. Appl. Math. Optim. **57**, 349–369 (2008)
56. Jasso-Fuentes, H., Prieto-Rumeau, T.: Constrained Markov decision processes with non-constant discount factor. J. Optim. Theory Appl. **202**, 897–931 (2024)
57. Jasso-Fuentes, H., Menaldi, J.-L., Prieto-Rumeau, T.: Discrete-time control with non-constant discount factor. Math. Methods Oper. Res. **92**, 377–399 (2020)
58. Kelley, J.L.: General Topology. Springer-Verlag, Berlin (1975)
59. Khan, M.A., Zaslavski, A.J.: On two classical turnpike results for the Robinson-Solow-Srinivisan (RSS) model. Adv. Math. Econ. **13**, 47–97 (2010)
60. Khlopin, D.V.: Necessity of vanishing shadow price in infinite horizon control problems. J. Dyn. Control Syst. **19**, 519–552 (2013)
61. Kolokoltsov, V.N.: Stochastic monotonicity and duality of one-dimensional Markov processes. Math. Notes **89**, 652–660 (2011)
62. Kolokoltsov, V.N.: Markov Processes, Semigroups and Generators. De Gruyter Studies in Mathematics, vol. 38. Walter de Gruyter & Co., Berlin (2011)
63. Kolokoltsov, V., Yang, W.: The turnpike theorems for Markov games. Dyn. Games Appl. **2**, 294–312 (2012)
64. Leizarowitz, A.: Infinite horizon autonomous systems with unbounded cost. Appl. Math. Optim. **13**, 19–43 (1985)
65. Leizarowitz, A.: Overtaking and almost-sure optimality for infinite horizon Markov Decision Processes. Math. Oper. Res. **21**, 158–181 (1996)
66. Leizarowitz, A.: An algorithm to identify and compute average optimal policies in multichain Markov Decision Process. Math. Oper. Res. **28**, 553–586 (2003)
67. Leizarowitz, A., Mizel, V.J.: One dimensional infinite horizon variational problems arising in continuum mechanics. Arch. Ration. Mech. Anal. **106**, 161–194 (1989)

68. Leizarowitz, A., Zaslavski, A.J.: Uniqueness and stability of optimal policies of finite state Markov Decision Processes. Math. Oper. Res. **32**, 156–167 (2007)
69. Makarov, V.L., Rubinov, A.M.: Mathematical Theory of Economic Dynamics and Equilibria. Springer-Verlag, New York (1977)
70. Marcus, M., Zaslavski, A.J.: On a class of second order variational problems with constraints. Israel J. Math., **111**, 1–28 (1999)
71. Marcus, M., Zaslavski, A.J.: The structure of extremals of a class of second order variational problems. Ann. Inst. H. Poincaré, Anal. non linéaire **16**, 593–629 (1999)
72. Marcus, M., Zaslavski, A.J.: The structure and limiting behavior of locally optimal minimizers. Ann. Inst. H. Poincaré, Anal. non linéaire **19**, 343–370 (2002)
73. McKenzie, L.W.: Turnpike theory. Econometrica **44**, 841–866 (1976)
74. Mordukhovich, B.S., Shvartsman, I.: Optimization and feedback control of constrained parabolic systems under uncertain perturbations. In: Optimal Control, Stabilization and Nonsmooth Analysis. Lecture Notes Control Inform. Sci., pp. 121–132. Springer, Berlin (2004)
75. Nikaido, H.: Convex Structures and Economic Theory. Academic Press, New York (1968)
76. Piunovskiy, A.: Turnpikes in finite Markov decision processes and random walk. Theory Probab. Appl. **68**(1), 123–149 (2023)
77. Piunovskiy, A., Zhang, Y.: Continuous-Time Markov Decision Processes. Probability Theory and Stochastic Modelling, vol. 97. Springer, Cham (2020)
78. Piunovskiy, A., Zhang, Y.: On the continuity of the projection mapping from strategic measures to occupation measures in absorbing Markov decision processes. Appl. Math. Optim. **89**, 25 pp. (2024)
79. Porretta, A., Zuazua, E.: Long time versus steady state optimal control. SIAM J. Control Optim. **51**, 4242–4273 (2013)
80. Prieto-Rumeau, T., Hernandez-Lerma, O.: Bias and overtaking equilibria for zero-sum continuous-time Markov games. Math. Methods Oper. Res. **61**, 437–454 (2005)
81. Radner, R.: Paths of economics growth that are optimal with regard only to final states: a turnpike theorem. Rev. Econ. Stud. **28**, 98–104 (1961)
82. Ramsey, F.P.: A mathematical theory of saving. Econ. J. **38**, 543–559 (1928)
83. Reich, S., Zaslavski, A.J.: Genericity in Nonlinear Analysis. Springer, New York (2014)
84. Rockafellar, R.T.: Convex Analysis. Princeton University Press, Princeton (1970)
85. Rotar, V.I.: Some retrospective remarks on pathwise asymptotic optimality. Dyn. Contin. Discrete Impulsive Syst. **19**, 207–224 (2012)
86. Samuelson, P.A.: A catenary turnpike theorem involving consumption and the golden rule. Am. Econ. Rev. **55**, 486–496 (1965)
87. Schweitzer, P.J.: On the solvability of Bellman's functional equation for Markov renewal programs. J. Math. Anal. Appl. **96**, 13–23 (1983)
88. Schweitzer, P.J.: A Brouwer fixed-point mapping approach to communicating Markov decision processes. J. Math. Anal. Appl. **123**, 117–130 (1987)
89. Shapiro, J.F.: Turnpike planning horizon for a Markovian decision model. Manag. Sci. **14**, 292–300 (1968)
90. Trelat, E., Zhang, C., Zuazua, E.: Steady-state and periodic exponential turnpike property for optimal control problems in Hilbert spaces. SIAM J. Control Optim. **56**, 1222–1252 (2018)
91. von Weizsacker, C.C.: Existence of optimal programs of accumulation for an infinite horizon. Rev. Econ. Stud. **32**, 85–104 (1965)
92. Zaslavski, A.J.: Ground states in Frenkel-Kontorova model. Math. USSR Izvestiya **29**, 323–354 (1987)
93. Zaslavski, A.J.: Turnpike Properties in the Calculus of Variations and Optimal Control. Springer, New York (2006)
94. Zaslavski, A.J.: Turnpike results for a discrete-time optimal control systems arising in economic dynamics. Nonlinear Anal. **67**, 2024–2049 (2007)
95. Zaslavski, A.J.: Optimization on Metric and Normed Spaces. Springer, New York (2010)

96. Zaslavski, A.J.: Stability of a turnpike phenomenon for a discrete-time optimal control systems. J. Optim. Theory Appl. **145**, 597–612 (2010)
97. Zaslavski, A.J.: Turnpike properties of approximate solutions for discrete-time control systems. Commun. Math. Anal. **11**, 36–45 (2011)
98. Zaslavski, A.J.: Nonconvex Optimal Control and Variational Problems. Springer Optimization and Its Applications. Springer, New York (2013)
99. Zaslavski, A.J.: Structure of Approximate Solutions of Optimal Control Problems. SpringerBriefs in Optimization. Springer, New York (2013)
100. Zaslavski, A.J.: Turnpike Phenomenon and Infinite Horizon Optimal Control. Springer Optimization and Its Applications. Springer, New York (2014)
101. Zaslavski, A.J.: Stability of the Turnpike Phenomenon in Discrete-Time Optimal Control Problems. SpringerBriefs in Optimization. Springer, New York (2014)
102. Zaslavski, A.J.: Turnpike Conditions in Infinite Dimensional Optimal Control. Springer Optimization and Its Applications. Springer, Berlin (2019)